THE ROUGH GUIDE to the

Best iPhone & iPad Apps

Peter Buckley

ROUGH
GUIDES

www.roughguides.com

Credits

The Rough Guide to the Best iPhone & iPad Apps

Text, design & layout: Peter Buckley
Editing: Caroline Parsons
Proofreading: Susanne Hillen

Rough Guides Credits

Publishing Director: Clare Currie
Publisher: Jo Kirby
Design Director: Scott Stickland
Production: Charlotte Cade

Apple hardware images courtesy of Apple UK

Acknowledgements

The author would like to thank everyone at Rough Guides.
Also, special thanks go to Duncan, Roz, Caroline and Rosalie.

Publishing information

This second edition published September 2013 by
Rough Guides Ltd, 80 Strand, London WC2R 0RL
Email: mail@roughguides.com

Distributed by the Penguin Group:
Penguin Group (USA), 375 Hudson Street, NY 10014, USA
Penguin Group (India), 11 Community Centre, Panchsheel Park, New Delhi 110017, India
Penguin Group (Australia), 250 Camberwell Road, Camberwell, Victoria 3124, Australia
Penguin Group (New Zealand), 67 Apollo Drive, Rosedale, Auckland 0632, New Zealand

Rough Guides is represented in Canada by Tourmaline Editions Inc.,
662 King Street West, Suite 304, Toronto, Ontario M5V 1M7

Printed and bound in Singapore by Toppan Security Printing Pte. Ltd.

1 3 5 7 9 8 6 4 2

Contents

Introduction

Why you need this book

There really does seem to be an app for every eventuality. And given the rate at which new apps appear and updates to your favourites roll out, it was pretty clear that we had to follow the success of this book's first edition with a fully revised second. So, whether you want to boil an egg perfectly at any altitude (see p.53) or record every step you take every day (see p.62) you'll find "an app for that" within these pages. If you can think of a need, then no doubt someone else has before you, and the app you're looking for is already live in the App Store. But there is a lot of rubbish out there, and reviews and ratings can be misleading – which is where this book comes in. The aim is to guide you to the best of the best and, hopefully, to introduce you to a few gems that you might otherwise have missed. And for the record, every app in this book is worth downloading.

About this book

This second edition was compiled in the spring of 2013, and all details were correct at the time of going to press. But it's worth noting that app developers are free to change their prices and features at any time, so apps that were once free may suddenly command a fee, and others might even disappear altogether. If you do find any discrepancies between our reviews and your downloads, please let us know (mail@roughguides.com), and equally, if you have any apps that you would like the world to hear about, please drop us a line.

It's also worth noting that apps sold through the iTunes App Store fall into three categories, according to which devices they will run on. Some apps are strictly iPad only, while others are designed for the smaller screened devices and are commonly known as iPhone apps (though they will also run on an iPod touch and an iPad, in the latter case with the same diminutive appearance as on the iPhone screen but with the option to double-up the pixels to fill the bigger screen). Finally, so-called "universal" apps will run on all iOS devices, with a specific user interface and set of functions tailored for each device's screen size. These apps are represented in this book like this:

■ **iPad app**

▢ **iPhone app**

▢ ■ **Separate iPhone and iPad apps available**

▦ **Universal app**

Note also that text written like **this** denotes a command or label as it appears on either a computer's screen or a device's screen. There are also some buttons and icons used within these pages, such as this ● and this ◉, which relate to buttons and icons as they appear on the iPhone or iPad's screen.

1
Managing apps
Downloading & organizing applications

There are tens of thousands of iOS apps out there, the vast majority of which can be downloaded inexpensively or for free. But before we get into the recommendations – the meat of this book – it's worth reading the following chapter so that you understand the best ways to download, store and organize your applications. Even if you're already an iPad or iPhone expert, we think you might pick up some useful tips and tricks from the next few pages.

Downloading apps

The iPhone and iPad both come with a bunch of apps pre-installed – from Mail and Safari to Stocks. But these are only the tip of the iceberg. To see what else is available, dive into the App Store.

Just as the iTunes Music Store changed music-buying habits, Apple's App Store – which provides apps for the iPhone, iPod touch and iPad – is quickly changing the way in which software is distributed. With so many apps available, the main problem is the potential for getting somewhat addicted and spending more time and money than you can afford. To use the App Store, you'll need the Apple ID username and password associated with your iTunes account.

Shopping with your Apple ID

Though anyone can browse the iTunes App Store, if you actually want to download something you need to set up an Apple ID and be logged in. If you haven't already done this, it's very easy: either try to buy something and follow the prompts, or head to **Settings > iTunes & App Stores** on your iPhone or iPad and enter the necessary details there. Here, you can also choose whether you want apps you purchase on other devices using the same account to be automatically downloaded to the device you're currently using and, more importantly, whether you want them to be downloaded over the cellular network (best to keep **Use Mobile Data** turned off if you have a capped data plan).

If someone else is already signed in to the Store on the same iPhone or iPad, they'll need to sign out first within **Settings > iTunes & App Stores**. Also note that iTunes App Store accounts are country-specific; in other words, you can only access the store of the country where the credit card associated with the account has a billing address.

Accessing the App Store

The App Store can be accessed in two ways:

• **On the iPhone or iPad** Simply click the App Store icon. Assuming you're online, you can either browse by category, search for something you know (or hope) exists, or take a look at what's new, popular or featured. When you find something you want, hit its price tag (or the word **free**) and follow the prompts to set it downloading.

Although most apps can be down-loaded via your cellular network, larger ones can take an eternity with anything less than a Wi-Fi connection.

• **On a Mac or PC** Open iTunes, click the **iTunes Store** button (top-right) and then hit **App Store** on the top nav bar. All the same apps are available and the interface for browsing them is, if anything, better than on either the iPhone or the iPad. Any apps you download in iTunes on your computer will be copied across to your iPhone or iPad the next time you sync. Or, if you have one Apple ID for iTunes and iCloud, you can access apps bought on any device by opening the App Store on your iPhone or iPad and looking in the **Purchased** list. Even better, with **Automatic Downloads** enabled (in **Settings > iTunes & App Stores** on your devices), apps you download via a Mac or PC will automatically appear on said devices almost instantaneously. Note that there are no refunds in the App Store, so it pays to read reviews before you buy.

> **TIP** Be careful if buying via iTunes that you choose apps that are suitable for your device – some are iPad-only, and won't work on the iPhone. Universal apps, suitable for both devices, are marked with a ✚ sign.

Buying apps with multiple accounts

Neither the iPhone, iPad nor iTunes on your computer have to be wedded to a single Apple ID. (You can have as many iOS devices as you like associated with a single Apple ID, but only five computers, Mac or PC running iTunes.) So if more than one member of your household uses the same iPhone, iPad, Mac or PC, there's no reason why you can't all have your own IDs and buy apps separately. Once installed on your iPhone or iPad, all the apps will be available to use, whichever account is currently logged into the device's App Store. However, to update an app, you'll need the username and password for the account through which it was purchased.

> **TIP** To log out of your account on your iOS device, tap **Settings > iTunes & App Stores** then tap the **Apple ID** field and choose **Sign Out**, or scroll to the bottom of most App Store windows and tap the **Apple ID** field there.

Updating apps

One of the best features of the App Store is that, as and when developers release updates for their software, you will automatically be informed of the update and given the option to install it for free, even if you had to shell out for the original download. To update apps:

• **On the iPhone or iPad** The number of available updates is displayed in a red badge on the corner of the App Store's icon. Tap the App Store app icon and then **Updates** followed by either **Update All** or an individual app's **Update** button to update one at a time. To pause the download of a specific update, tap its app icon on the Home Screen while it's displaying a progress status bar.

> **TIP** If you receive an error message when tapping **Update All** claiming that you do not have enough free space on your device to continue, update the apps one by one instead. If the problem continues, try freeing up some space by deleting infrequently used apps (see p.18).

• **On a Mac or PC** The number of available updates is displayed on a button at the bottom of the iTunes window when **Apps** is selected in the **Library** switcher button, top-left. Click this **Available Updates** button to either download all or individual updates.

Fast app switching

If your iPhone or iPad is running iOS 4 or later, then double-
clicking the Home button when
your device is in use reveals the
Multitasking Bar, where your
recently used applications are displayed. This is extremely handy,
enabling you to quickly switch
between apps without having to
return to the Home Screen.

Swiping to the left across this
Multitasking Bar reveals the next few most recently used apps, and
so on. Tapping an app's icon swiftly switches you into that app, and
for most apps you'll find that you are back in exactly the same place
you were last time you used it.

Should you want to remove apps from the Multitasking Bar to
restart them properly, tap and hold one of the icons until they all
start to wobble; then tap the red ⊖ icon of any apps you want rid
of. Click the Home button when you're done. This process does not
remove the app from the device altogether, only from this list.

TIP The iPhone and iPad allow multitasking, which means
apps can perform tasks in the background while you
get on with something else in another app. However,
Apple has limited the types of features that can run in the
background in order to preserve battery life and stop your
device grinding to a halt. Background functions supported
include music playback, receiving VoIP calls and location
awareness.

Switching to the current music playback app

Swiping to the right across the Multitasking Bar reveals music controls and an icon that gives you quick access to whichever app you are using for music playback: for example Apple's own Music app, Spotify (see p.78) or Bloom (p.79).

To the left of these playback controls there's a special button for locking the iPad or iPhone's screen orientation – handy when you're reading or using various apps in bed. When the screen orientation is locked, this button displays a padlock in its centre.

Web apps & web-clips

Downloading apps is one way to fill up your Home Screens with icons; the other is to create simple bookmark icons – so-called "web-clips" – that point to your favourite websites or web apps (see box overleaf). These can be handy as they allow you to access your most frequently viewed websites without opening Safari and manually tapping in an address or searching through your bookmarks.

To create an icon for a website, simply visit the page in Safari and press the ➦ icon. Select **Add to Home Screen** and choose a name for the icon – the shorter the better, as anything longer than around ten characters won't display in full. (For more on webclip Home Screen icons, see p.15.)

Web apps and optimized sites

Whereas a "proper" iPhone or iPad app is downloaded to your iOS device and runs as a standalone piece of software, a web app is an application that takes the form of an interactive webpage. The term is also often used to describe plain old webpages that have been specially designed to fit on a mobile device's screen without the need for zooming in and out. You might also hear people refer to them as "iPhone-optimized" webpages and websites, or sometimes "mobile" webpages or websites.

Unlike many proper apps, web apps will usually only work when you're online. On the plus side, they tend to be free – and they usually take up no space in your iPad or iPhone's memory. There are exceptions to these rules, however. For example, Google's Gmail web app uses a "database" or "super cookie" to store some information locally – which does take up some space but enables a few functions to be carried out even when offline.

Several websites – including Apple's own – offer directories of iPhone-optimized websites and web apps, arranged into categories. Though many of these sites look out of date, due to the rise of fully fledged iPhone and iPad apps, there is a new wave of web apps appearing that are rich in features and employ HTML5 technology to create a very app-like experience.

Apple's web app directory apple.com/webapps

Whether or not they call them "web apps", many popular websites offer a smaller version for browsing on mobile devices in general or iPhones specifically. These not only fit nicely on small screens but also load much faster, since they come without the large graphics and other bells and whistles often found on the full version of their websites.

Usually, if the server recognizes that you're using a mobile web browser, the mobile version of a site will appear automatically. Occasionally, though, you'll have to navigate to the mobile version manually – look for a link on the homepage. Mobile sites often have the same address but with "m", "mobile" or "iphone" after a slash or in place of the "www". For example:

Digg digg.com/iphone
eBay mobile.ebay.co.uk
Facebook m.facebook.com
Flickr m.flickr.com
MySpace m.myspace.com

Improving website icons

When you add a webpage to your Home Screen, the iPhone or iPad will display a custom logo icon for that website, if one has been specified. If the site doesn't have an available icon, an icon will be created based on how the page was displayed when you clicked ⤴. To get the best-looking and most readable icons, zoom in on the logo of the website in question before clicking ⤴.

Organizing your apps

As your iOS device fills up with apps and web-clips, they'll soon start to spill over onto multiple screens, which you can easily switch between with the flick of a finger – simply slide left or right.

The number of screens currently in use is shown by the row of dots along the bottom of the screen, with the dot representing the current screen highlighted in white. You can return to the Home Screen by pressing the Home button or by sliding left until you reach it.

Moving icons

To rearrange the apps and webclips on your screen, simply touch any icon for a few seconds until all the icons start to wobble. You can now drag any icon into a new position, including onto the Dock at the bottom of the screen (to put a new icon here, first drag one of the existing ones out of the way to clear a space).

To drag an icon onto a different screen, drag it to the right- or left-hand side of the screen. (To create a new screen, simply drag an icon to the right-hand edge of the last existing screen.) Once everything is laid out how you want it, click the Home button to fix it all in place.

App folders

Instead of just having your apps and links arranged across various screens, it's possible to group relevant apps together into folders. You might create, for example, a "travel" folder containing a mix of map apps, guidebooks and links to travel-related websites.

You could simply group related icons together onto specific screens, but folders offer some advantages, especially if you've got loads of apps. For example, they allow more apps to live on your Home Screen, making them more readily accessible, and generally keep things tidier.

To create a folder, first touch any icon for a few seconds until they all start wobbling. Then simply drag one icon onto another, and a folder will be created containing both icons. iPhone folders can contain up to twelve apps, while iPad folders can hold twenty. Your iPhone or iPad will attempt to give your new folder a name based on the categories of apps you've combined – tap into the title field to overwrite this with a name of your choice. To add another app to the folder, simply drag it onto the folder's icon. To remove an app, click

Checking app settings

Just like the apps that come pre-installed, many third-party apps have various preferences and settings available. Many people overlook such options and can end up missing out on certain features as a result. Each app does its own thing, but expect to find some settings either:

• **In the app** If there is nothing obvious labelled **Options** or **Settings**, look for a cog icon, or perhaps something buried within a **More** menu.

• **In the Settings app** Tap **Settings** on the Home Screen and scroll down to see if your app has a listing in the lower section of the screen. Tap to see what options are available.

on the folder and drag the relevant app out, either into another folder or back onto the Home Screen.

> **TIP** Once you've got lots of icons, you may find it easier to organize them into screens and folders in iTunes, which allows you to see multiple screens at once and to move icons around with a mouse. To do this, connect your iPhone or iPad to your computer, select its icon in iTunes and click the **Apps** tab.

Deleting apps

It's worth noting that your Apple ID account keeps a permanent record of which apps you have downloaded, so if you do delete both your device's copy and your backed-up iTunes copy, you can go back to the Store and download it again at no extra charge.

You can't delete any of the built-in apps, but you can move them out of the way onto a separate screen or into a separate folder if you don't use them.

• **Deleting apps and web-clips from Home Screens and folders** Hold down any app icon until they all start wobbling and then tap the black ⊗ icon on the app you want to delete. When you have finished, hit the Home button. This will not remove the app from iTunes or iCloud, so you can always sync it back again later.

• **Deleting apps from Settings > General > Usage** Here you'll find a list of all your installed apps arranged in size order. Tap any of the listed apps to find the **Delete App**

option. This can be very useful if you need to make room on your device quickly for new content or downloads.

• **Deleting apps from a Mac or PC** Highlight **Apps** in the iTunes sidebar, then right-click the unwanted app and choose **Delete** from the menu that appears. If the app was also on your iPad or iPhone, the next time you connect you'll get a prompt to either remove it from there too or to copy it back into iTunes.

2
Books

Readers, comics & kids' books

In this chapter, we'll take a look at the best apps for getting a slice of literature onto your iPad or iPhone, as Apple's own iBooks app isn't the only way to read books on your device. When it comes to specialist graphic titles such as comics and manga, for example, there is an amazing selection available in the App Store, where many publishers produce their own free "store" or "library" apps, with individual titles available to buy using your regular Apple ID from within the app (these are known as in-app purchases).

eBook reading apps

iBooks

 Free

Apple's iBooks is not one of the default apps that comes pre-loaded on the iPhone or iPad. It is a free app, but you have to visit the App Store to download it. The reading experience in iBooks is great for both fiction (you can change the font size and read in "Night" mode) and for more graphic textbook-style titles (publishers such as DK are now producing stunning reference books that feature animation, video content and 3D models).

The iBooks app also offers tools for adding notes, bookmarks and highlights, and can be used to store and read PDF files as well as eBooks downloaded from Apple. You can also share quotes from books via Facebook, Twitter, SMS or email.

> **TIP** If you're sent a PDF as an email attachment and view it in the Mail application, look out for the option behind the ☑ button to save the file to your iBooks library.

Unfortunately, not every country has an iBookstore, so if you can't find the iBooks app in your local App Store you're going to have to find another way of getting books onto your iPhone. Read on.

Kindle

 Free

For reading Amazon eBooks this is a great little app, and it has a similar set of features to iBooks. You can't actually browse the Amazon store from within the app; for that you need to head to the website and then sync your purchases over the airwaves. For reading PDFs and other documents within the app, Amazon give you a special email-to-Kindle email address that lets you send files straight to the app as attachments.

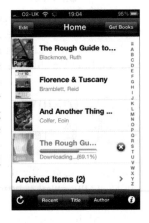

Stanza

 Free

This reader, and its associated store, offer thousands of titles, including plenty of free classics from Project Gutenberg, and you can also add your own files using the Stanza Desktop application.

Google Play Books

Free

Syncs via the cloud with your Google Books account and has a nice reader interface which, like Stanza and iBooks, includes a white-on-black night-reading mode. As with the Kindle app, you have to shop for titles outside of the app on the associated website.

Kobo

 Free

A nice-looking app that gives you access to thousands of titles from the Kobo store. The reading experience is excellent, with font and bookmarking options. There is also a useful feature that lets you add titles via either an iCloud or Dropbox account.

Inkling

 Free

One of the more interesting new kids on the block, the Inkling app opens the door on a unique reading platform, where books are presented as browsable sets of cards. With more and more publishers making their titles available through the Inkling store, it's well worth investigating.

Kids' reading apps

There is a glut of kids' content to be found in the App Store Books category, but do read the reviews and be selective, as the majority of what's there is shockingly poor. Here are a few that are worth a go:

The Gingerbread Man

 Paid

One of many read-along titles, with integrated audio, from the Kidztory stable. These books really bring the tales to life, and there are surprises on each page to keep tapping fingers happy.

Ladybird Classic Me Books

📱 Free (+in-app purchase)

This beautiful reader app gives you access to the iconic Ladybird series. You get *The Zoo* for free, and then build your collection via in-app purchases. Many feature audio read by famous names, and you can even record your own sound effects and narrative as you read. There is also a separate Me Books app that gives you access to a range of pre-school titles from other publishers.

> **TIP:** To stop your kids racking up a massive bill when using apps with in-app purchases, look to **Settings > General > Restrictions** to disable the **In-App Purchase** function.

Nursery Rhymes with StoryTime

📱 Paid

One of a growing number of apps that strive to take the book/app hybrid format to the next level. With audio, moveable objects and delightful graphics, this is a treat for all ages and features loads of classic nursery rhymes. And best of all, mums and dads away from home can use the app to remotely read to their little ones.

Dr. Seuss Bookshelf

📱 Free (+in-app purchase)

Perennial kids' favourites ingeniously adapted for the screen. You can read it like a book, listen to the narrative or play the whole thing through like a movie.

Comics and graphic novels

Marvel Comics

Free (+in-app purchase)

Get access to an essential store of classic superhero comics – *Spider-Man*, *Thor*, *Wolverine* and *Iron Man* are all here – alongside a seemingly endless list of more obscure Marvel material just waiting to be discovered. Download and read either page by page (as you would in a regular comic) or double-tap to read frame by frame. If you also have a Marvel.com account you can back up your purchases, making it easier to keep track of your library.

Comics

Free (+in-app purchase)

Used in conjunction with a ComiXology account this app really delivers both in terms of features and content. Alongside all your favourites from Marvel, IDW, DC and the like, there are extensive lists from Disney, making it a good option for comic-crazy parents with kids who might want something to read too.

Manga Storm

Lite/Paid

App offering access to five of the most important Japanese manga websites. New chapters are automatically delivered to you when they become available. With the free app you can bookmark three mangas, but you'll need the paid version for more.

IDW Comics

Free (+in-app purchase)

The complete IDW collection available in one app. Like the other stores, you buy individual comics and then organize your own library. The same publishers also produce dedicated series comic apps for *Transformers*, *Doctor Who*, *Star Trek*, *True Blood* and many more.

TIP As well as all the "store" and "library" apps, thousands of comics can also be found in the store published as standalone apps.

Miscellaneous

Audible

📱 Free

The iPhone's Music app does a perfectly good job of playing audiobook content purchased from iTunes, but if you want a dedicated audiobook app, this is the one to go for.

Bookscan for iBooks

📱 Free

Point your iPhone's camera at the barcode on any printed book, and this app will tell you whether the title is available in the iBookstore and present you with a link to get there.

Brilliant Quotes & Quotations

📱 Free

Though you can find similar titles in the App Store Reference section, this is one of the better quote collections.

The Silent History

📱 Paid (+in-app purchase)

A special mention goes out to this interactive eBook experience … one of the more innovative to appear in 2012. The initial price gives you the first chapter, which will have you hooked in minutes.

3

Business & Finance

Apps for city slickers

The Business and Finance categories of the App Store are peculiar places. Of all the areas of the Store, these are the most littered with unlikely apps, many of which have little or nothing to do with either finance or business. That said, with a little digging, there are some real finds here, especially if you trade in stocks and shares, need to keep detailed financial records or run a small business.

Stocks and shares

Stocks

📱 **Built-in**

The iPhone's built-in Stocks app (you won't find it on the iPad) lets you view current values for any listed company. Well, not quite current – the prices are refreshed whenever you open the application, but are typically still about twenty minutes out of date.

> **TIP** The stocks you choose will also appear in Notification Center, which you can reach by swiping down from the very top of the iPad and iPhone Home Screen. To disable this ticker, visit **Settings > Notifications > Stock Widget**.

MarketDash

📱 **Free**

A nicely designed iPad app for covering your favourite stocks, shares and breaking market news. Sync with a Yahoo! Finance online account to manage a complete portfolio.

Thomson Reuters Marketboard

📱 **Free**

This app is another good punt for the iPad; it has a mix of charts and tools for following your chosen companies alongside news and analysis. It also has an offline "Briefcase" mode, which lets you pull together content to sift through when you're not connected.

OANDA fxTrade Forex Trading

📱 Free

Oanda is already known to many in business as one of the leading internet-based services for currency trading and data, so it's no surprise that they have a very serviceable app in the App Store. Not too many bells and whistles, but it does a great job of making on-the-move trading quick and easy.

Currency converters

XE Currency

📱 Free

XE Currency is arguably the best currency converter tool out there. It really comes into its own when you're abroad and need to quickly check whether or not something is a bargain. It also, usefully, lets you compare multiple currencies simultaneously.

Currency Converter

📱 Free

This application offers more sophisticated comparisons between currencies. You can use the Interbank Exchange Rate, or, alternatively, calculate your conversions taking into account the charges you're likely to accrue from your chosen currency exchanger.

Personal finance

Balance Guide

📱 Paid

Use this stylish tool to keep track of your bank account balances. Once it's set up, you can enter recurring transactions (mortgage payments, for example) so that you can more accurately predict where you're likely to be in the near future, as well as how things are looking at any given moment.

Money Smart (HD)

📱 📱 Lite/Paid

One of many budgeting apps that help you keep an eye on your spending. You enter your budget and then document all your spending within different categories. You can also create graphs to see your spending patterns. The Lite version limits the number of transactions you can record.

Cube Time & Expense Tracker

📱 Lite/Paid (+in-app purchase)

Targeted at professionals, this app hinges on the idea of tracking your time and expenses. It's easy to set up, and while the free version is good enough for basic tracking, the Pro version allows you to export data and to receive notifications when your records need updating. And for tracking expenses of a group and then dividing up the total (perhaps while on holiday with friends), try the excellent FriendCash universal app.

Mint.com Personal Finance

📱 Free

If you're one of the many people who already use the excellent Mint.com money management tools, then you'll want to check out their free app too.

DailyCost

📱 Paid

A good-looking iPhone app for tracking your daily outgoings. Held in portrait mode you manage your expense items by category and switch to landscape to see statistics and graphs.

OrSaveIt

📱 Paid

Like having a little conscience sitting there on your phone, this app encourages you to regularly forgo life's little luxuries and instead put money aside for a bigger treat.

Still Waitin

📱 Paid

Create automated reminders to help keep track of cash that you have lent to friends.

Small businesses

Employment Law Cloud

📱 **Free**

Aimed at UK employers (in any size business), this app is a first-class reference tool for all those nagging HR questions. Also check out the excellent universal app Fingertip HR.

Easy Books

📱 **Free (+in-app purchase)**

Accountancy made easy for the iPad and iPhone. This app offers a full double-entry accounting system that can be used as both a record-keeping tool and a reporting tool.

Shorthand Manager

📱 **Free**

Putting to one side its slightly misleading moniker, this is a great little app for organizing calendars and tasks in a small office team. Alongside scheduling tools, there's also a handy marketing screen for sending group emails or posting a social media update.

HotSchedules

📱 **Paid**

Used in conjunction with a HotSchedules online account, this app is great for managing shifts within a team.

Direct Report

🔲 🔲 Paid

With two separate apps for the iPad and iPhone, Direct Report offers a useful set of tools for monitoring and recording employees' development and feedback.

SG Project

🔲 Paid

This exceptional app is all you need for both small and larger-scale project management. The Gantt chart-style view is great and the Dropbox support allows you to share project schedules with ease. There is also an iPhone version called SG Project Go and a Mac version, which makes syncing your projects very straightforward.

Quick Sale Lite

🔲 🔲 Free (+in-app purchase)

This is a very nice tool for creating, recording and dispatching professional-looking invoices. Under the hood you'll also find a selection of inventory tools and support for various iPad wireless printing systems.

Invoice2go

🔲 Paid (+in-app purchase)

Another excellent invoicing and reporting tool. This one has a plethora of additional functions available via in-app purchase.

Point of Sale

Free

This free app is the preferred choice of many small retail businesses that have switched to the iPad for recording their sales during the course of the day. It has an easy-to-use interface, including the ability to quickly and easily create new stock items, and because it's on an iPad, it's perfect for taking out and about to trade shows and fairs.

Square Register

Free

You might be interested in the idea of collecting card payments via your iPhone – especially if your business is conducted door-to-door or out and about. This app and service from Square is currently only available in the US, but it's sure to be the kind of thing we see a lot more of in coming years; visit squareup.com to sign up and get hold of the card-reader hardware. The same company also make a Square Wallet app for making payments at participating outlets.

4

Education
Learning made easy

Though we are very much at the beginning of the technological revolution when it comes to education in the classroom, education at home, especially for the very young, is bounding ahead as more and more touch-screen devices are becoming commonplace in living rooms across the world. In this chapter we look at some of the apps that are helping both school and college students get organized and educated. We also take a whistle-stop tour of the best-in-show educational apps for younger brains, with a particular focus on apps that blur the boundaries between learning and play.

Higher education

If you're a student heading off to university or college, it's worth checking in the App Store whether your institution has any apps available and what services they offer. Many include maps, time-tables, event listings, library reminders and the like, while others might even give you access to your grades and tutorial feedback.

iTunes U

📱 **Free**

iTunes U is Apple's fully featured education portal, giving you access to iBooks textbooks, video lectures, podcasts, presentations and even complete courses created and presented by academic professionals from across the globe.

iStudiez

📱 **Lite/Paid**

A dedicated calendar and organizer, with tools to help you keep track of your grade averages. With the paid-for Pro version of the app you also get to save backups and receive push notification alerts that should help you stay on track.

Class Timetable

📱 **Lite/Paid**

Much simpler than iStudiez, this is the best of the free, feature-light timetable apps in the store. The Pro version is available as an in-app purchase and has a nice calendar view.

PhotoFit Me

Free

Create a photo-fit image of yourself to understand the way the human brain recognizes faces. Check out the OU's equally entertaining evolution-tinged Devolve Me app too.

School days

iHomework

Paid

There's lots going on in this organization tool: calendars, alerts, coursework schedules, teacher contacts lists and more. A desktop version is also available to help you keep everything in sync.

Maths

Free (+in-app purchase)

This library of around four hundred maths tutorials and refreshers is great for both cramming and homework. The first few lessons are free, but you'll need to sign up to get unlimited access.

Autodesk Digital STEAM

Free

The basics of applied dynamics presented through play. The games lead to applied problems to test what's been learnt.

Educational games

Baby Flash Cards – Encyclopedia 300+

Lite/Paid

There are hundreds of flash cards apps in the Store, and most aren't worth the time it takes to download them. This app, however, makes good use of both graphics and photos. The synthesized voice isn't great, but on balance this app is a good option.

Bugs and Buttons

Paid

An exceptional app for kids. These eighteen beautifully rendered games really strike the balance between education and fun. The activities intelligently tackle everything from counting and sorting to motor skills and pattern recognition. Not, however, recommended for little ones with creepy crawly phobias. The same developer also makes the excellent Bugs and Numbers.

Hickory Dickory Dock

Paid

It's time to learn to read a clock, and there really isn't a nicer way to do it than with this impressive app that, as you might expect, hangs the learning experience off the familiar nursery rhyme. The same company also make the excellent word-matching Opposites app for a slightly older age group (7+).

Toca Kitchen Monsters

 Paid

Stressing the importance of "free play", this educational cooking app lets children explore the world of food in a novel way. Combine different ingredients and cooking methods to see what comes out. The Toca Boca company also make the excellent Toca Tailor, Toca Train and the rockin' Toca Band.

My First Animal Games

 Paid

This sequel to DK's My First Words app is another great introduction to reading for the very young, with loads of letter-based games, a free-play animal sticker board, a "mothers & babies" matching pairs game and a gorgeously rendered paint-and-reveal feature.

abc Pocket Phonics

 Lite/Paid

Designed and tested by teachers from the UK, this app is a great introduction to phonics (letter sounds) for pre-schoolers.

From Earth to Space

 Paid

This app beautifully explores subjects ranging from the Earth's core to orbiting satellites. There's a built-in quiz so that your kids can test themselves as they learn.

Numberlys

 Paid

This app brings you educational storytelling at its best. Unlike anything else in the App Store right now, this app presents – in glorious black and white (which can be a hard sell for some kids) – the story of the alphabet.

DK Kids' Craft

Paid

This app, based on the popular Jane Bull books, is ideal for those rainy days during the school holidays. The in-app craft tools are delightful, with virtual cross stitch and furry felts being the big-hitters in my household. There are also detailed instructions for "real world" craft projects, such as knitting a bear and stitching a pirate puppet.

Kids' Vocab - MindSnacks

 Free (+in-app purchase)

A chirpy app of vocab games that look to develop phrasing skills as well as individual word skills.

The Land of Me: Story Time

 Free

Created by child development experts, this charming tale, with its adaptive storyline, is well worth downloading.

5

Entertainment

TV, video & the movies

The Entertainment category is probably the most eclectic in the App Store. It contains a mix of apps that relate to the world of entertainment (TV and movie listings and streaming services, for example) plus countless little joke apps that are striving simply to entertain (pranks, talking chickens and the like). As you might imagine, in the case of the latter, the term "entertainment" is often pushed to the bounds of its definition, making it very easy to spend quite a lot of cash on apps that are about as entertaining as food poisoning – so tread carefully. This chapter concentrates on the former definition of entertainment, with a focus on the TV streaming and video-player services that really come into their own when used on the iPad and iPhone. And don't forget to get your paws on the YouTube app – once installed as a pre-load on iOS devices, it now has to be downloaded from the Store.

Streaming live & catch-up TV

The main limitations to watching either live or catch-up TV on the iPhone and iPad are regional: many services that are currently accessible in Europe aren't in the US, and vice versa. You are sure to find some that work, but don't be too surprised if a few of the apps listed below dish you up a whole lot of nothing.

BBC iPlayer

 Free

This is an essential app for catch-up (and live) TV and radio streaming. The app works over both 3G and Wi-Fi connections, though do keep your data plan in mind if you get addicted to the cellular feed. The BBC also have a separate iPlayer Radio app with a nice dial-based interface.

4oD Catch Up

 Free

The UK's Channel 4 also offers an essential app for catch-up streamed viewing. It boasts a nice selection of content, and there are useful parental control settings too.

Netflix

 Free (+subscription)

Sign up for the Netflix service and get unlimited access to their library of movies and TV series. There's plenty of great stuff, but the new release selection isn't always as current as you might like.

Hulu Plus

📱 📱 Free (+subscription)

This app gives you access to the popular US subscription service. Loads of the major networks are covered, with season passes, 720p HD offerings and tools for managing your play queue.

Boxee

📱 Free (+subscription)

Boxee are one of the big names in web-based streaming TV services and set-top-box systems. At the time of writing, the Boxee app is iPad-only, but it's rumoured that an iPhone version is in the pipeline. As well as giving you access to a host of online channels, this app will also let you stream any video format from your networked desktop computer to your iPad – particularly handy when you've got formats that won't otherwise work with iTunes and the iPad's Videos app. For the latest news about the Boxee service, visit boxee.tv.

TVCatchup

Free

This app is the only one you need to stream live UK Freeview TV to your iPhone or iPad (though you do have to put up with an advert at the start of each session). It isn't great over a cellular data connection, but works very well over Wi-Fi.

Sprint TV for iOS

Free

The basic Sprint TV streaming service (US-only) is free with many phone data plans, so it's worth finding out if that applies to you. That said, beware the premium paid-for selections, as they're charged to your monthly phone bill, making it easy to lose track of exactly how much you've spent.

Other video players

The biggest limitations of the iPhone and iPad, when it comes to video playback, are the formats that are supported by the Videos app. Thankfully, several apps offer an alternative:

GoodPlayer

Paid

Great all-round video player for the iPhone and iPad that can handle pretty much any file format you throw at it (Xvid, Divx, MKV, MP4, etc).

OPlayer & OPlayerHD

▢ ▢ Lite/Paid

Another good option that can handle all manner of files. OPlayer also has loads of additional features and tools for organizing and playing your files. Movies can be sideloaded via iTunes or transferred to the app via Wi-Fi syncing of ftp.

Plex

▣ Paid

If you use the excellent Plex media server service to make your music, video and photos available for personal use across the web, then check out their iOS app to get access from your iPad or iPhone. To get the system up and running you'll also need the Plex Media Server running on your desktop machine at home. Find out more at www.plexapp.com.

Recording from TV

If you want to create iPad- or iPhone-friendly videos by recording from television, your best bet is to use a TV receiver for your computer. Hauppauge, Freecom and others produce USB products for **PCs**. Browse Amazon or other major technology retailers to see what's on offer.

The obvious choice for **Mac** users is Elgato's superb EyeTV range of portable TV receivers, some of which are as small as a box of matches. You can either connect one to a proper TV aerial or, in areas of strong signal, just attach the tiny aerial that comes with the device. The most interesting hardware for app fanatics are the Netstream DTT and Netstream Sat TV tuners, which both connect wirelessly to the Elgato EyeTV Netstream app. For more info, see: elgato.com

Remote controls

Remote

Free

Free to download from the App Store, Apple's Remote app can be used to control iTunes on your computer via Wi-Fi. Coupled with an Apple Airport Express unit, this can be a great way to stream music and video around your home.

Rowmote Pro

Paid

Although it doesn't come free, this super-charged remote control app is well worth having a play with, as it gives you control of all sorts of applications on your Mac computer (such as Hulu Desktop and Boxee), not just iTunes. It can also be used for controlling presentations from both PowerPoint and Keynote.

VLC Remote

Lite/Paid

If you use the VLC player on your computer for watching movies, this remote control app is a welcome addition. Also check out its sister universal app VLC Streamer, which can be used to view video files located on your computer on an iPad or iPhone.

TIP There are also loads of third-party remotes in the App Store for controlling the likes of Spotify and Boxee.

TouchPad

 Paid

Best of many apps on the store for remotely controlling your Mac or PC from an iPhone or iPad.

Listings & info

TV Guide Mobile

Free

One of many TV-guide apps that can be found in the Store, this US app displays channels based on your ZIP code and provider settings. In the UK, a good equivalent is simply called TV Guide, and it comes in separate iPad and iPhone versions.

IMDb

 Free

This popular online source for movie and TV information is now in app form – an absolutely essential download.

Flixster

 Free

A one-stop-shop for all your movie needs: local show-times, trailers, upcoming movies, new DVD releases, reviews from Rotten Tomatoes and more.

Amusements

Seemingly every week a new iPhone amusement app is doing the rounds; it doesn't happen so much with iPad apps, as it's really all about the sharing process ("Hey, have you seen this?"), which is much easier with a phone than a tablet. Here are a few of our favourites from recent times:

Drawing With Carl

Paid (+in-app purchase)

Unless you have been living on the moon for the last few years, you have no doubt experienced Talking Carl. The app has a simple concept: you speak to him, and he talks back in a squeaky voice. Drawing With Carl finds our hero helping you to get to grips with art. There are tons of tools, brushes and expansion packs.

Talking Tom Cat 2

Free (+in-app purchase)

The original Talking Tom Cat (a variation on the Talking Carl theme) was an App Store sensation from day one. It has since spawned a talking toy, a sister app called Talking Ginger, and this feature-rich sequel (in which you can shop for cat accessories).

Volt

Paid

Use this app to fool people into thinking they have received an electric shock from your iPhone or iPad's screen.

BannerFlo

📱 **Free**

Create animated banners that scroll across your iPhone's screen. Use the clock mode to display the time.

Bubble Wrap

📱 **Lite/Paid**

Virtual bubble wrap for popping – who would have thought someone could make a fortune selling it? Well, they have!

MouthOff

📱 **Lite/Paid**

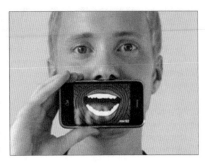

Simply hold your iPhone in front of your face, select one of the many amusing mouths on offer, and start nattering. It's really stupid, but also immensely funny.

Tattoo You

📱 **Free (+in-app purchase)**

Ever wondered what you might look like with a tattoo? Maybe this app can help. The in-app purchases get you additional tattoo packs to plaster all over your favourite photos.

How to make Origami

📱 Free

Wonderful app with step-by-step, 3D, animated instructions for crafting beautiful origami creations.

ZX Plectrum

📱 Paid

Touch the screen and this app recreates the load sounds and garish graphic static of the vintage ZX Spectrum computer.

Shopping for apps

The App Store charts are a great place to find inspiration for what to download next, but there are also many excellent apps dedicated to helping you find the best apps around, plus they offer personalized settings and search options. Check these out:

AppoDay 📱 Free

Every day this tool recommends an app that is currently free to download, and then nudges you to rate and review it.

AppShopper 📱 Free

This product features highly customizable settings to hone its app recommendations, and also delivers notifications and alerts when an app you're interested in is on special offer or has its price dropped.

Discovr Apps 📱 Free

This really nice app (pictured) offers recommendations based on the apps you've already bought and rated.

6

Food & Drink

**Recipes, kitchen utilities
& food-finders**

This sprawling category within the App Store is predominantly made up of recipe apps, representing every major cuisine and every major celebrity chef. While most of these apps contain delicious edibles, their usability varies wildly ... nowhere is the use of a touch screen more challenged than in the kitchen when fingers are sticky. Look for recipe apps that offer clear step-by-step instructions designed to be used while actually cooking. And don't forget to also checkout the titles in the iBookstore where many more recipe books are too be found.

Recipes

Macgourmet touch

 Paid

This app is the sister offering of the Macgourmet desktop Mac application, which lets you aggregate recipes from multiple sources in one place. The iPad app provides a special "chef view" with step-by-step instructions as you create your culinary masterpieces.

Epicurious Recipes

Free (+in-app purchase)

With access to a massive database of recipes, this is among the best cooking apps available. You can also use the app to generate shopping lists and read recipes in a step-by-step mode.

DK Quick Cook

Paid (+in-app purchase)

This app's well-presented set of recipes is easy to follow. The recipe filters let you choose what you could cook based on your available ingredients, and how much time you have available.

iBartender

Paid

This collection of cocktail recipes allows you to search either by name or by a list of the liqueurs you have to hand.

Kitchen helpers

Escoffier Cook's Companion

 Paid

With a built-in measurements convertor, equipment list, timer and glossary of terms, this very well designed app is sure to come in handy while cooking.

Gourmet Egg Timer

Paid

There are some wonderful kitchen gadgets waiting to be found in the App Store. For the perfect boiled egg, this app takes into account altitude, egg size and how soft you like your yolk. You can also have multiple timers running simultaneously if family members like their eggs done differently.

Substitutions

 Paid

This is a simple but brilliant app that tells you what to use in place of that key ingredient that you've run out of.

Freshbox

 Paid

Record the shelf life of your perishable groceries and get reminders to use them before it's too late.

Food-finders

Foraging

📱 **Paid**

Foraging for wild food is all the rage at the moment. This beginner's guide will show you how to get stuck in and also record the locations where you came across your wild bounty.

AnyList

📱 **Free**

This fully featured shopping list app lets you easily create and cross off items, manage collections of recipes and their ingredients and also share your lists with a spouse or friends.

Evernote Food

📱 **Free**

What greater pleasure is there in life than eating? How about photographing and journaling what you eat? This app lets you organize your culinary adventures in multiple ways and then sync it all with your Evernote cloud account.

Foodspotting

📱 **Paid**

For a more travel-oriented take on food journaling and dish reviewing, try the excellent Foodspotting app.

7

Games

Killing time on the iPhone & iPad

Both iPhones and iPods before them originally came with a smattering of games built in. But with the arrival of the App Store, gaming on Apple hardware became much more than just light distraction. Playing games on iPhones, iPods and iPads is so popular that whole new brands have grown up around it – the must-have Angry Birds being an obvious example. Today, over fifty percent of the top-ranking apps in the Store at any one time are games. And it's not just traditional action-packed videogames – blood-splattered zombies and high-octane racing – that are being consumed in vast quantities (though if that's your vice, you won't be disappointed). In this chapter, we'll make suggestions on everything from fiendishly challenging puzzles to table tennis.

Puzzles, quizzes & strategy

Chess Free (HD)

📱 📱 **Free**

Dozens of chess apps are available in the App Store. This one is free, looks nice and does everything you need it to do.

4 Pics 1 Word

📱 **Free (+in-app purchase)**

This simple, but often cunningly tricky quiz has you guessing the word that links four images. After only a few months in the store there were dozens of clones, and also cheat apps.

Osmos

📱 📱 **Paid**

A mellow cosmic adventure in which you battle gravity and mass to "become the biggest" – particularly good on the iPad.

DK Quiz

📱 **Free (+in-app purchase)**

Based on the popular DK Quiz website, this app tests your knowledge across dozens of categories and subject areas, with quizzes rated by difficulty. Once you have played a quiz you can challenge friends and family via Facebook to better your score. Use the in-app purchases to get additional coins to buy lifelines.

Rubik's Cube

 Paid

A virtual version of the classic cube, which can even be played in 3D mode using a pair of 3D glasses. Additionally, you can use this app to take photos of a real Rubik's cube and get hints on how to solve it.

Cut The Rope (HD)

 Paid

A beautifully rendered puzzle game that has you feeding candy to a devilish little green critter.

Logos Quiz Game

 Free (+in-app purchase)

Test your knowledge of logo graphics by identifying graphics with key elements removed. Of the many variations on the quizzing theme to be found in the App Store, this app (like Draw Something before it) really captured the public imagination.

Bejeweled

Paid

Accept no imitations, this PopCap classic is a game that you'll come back to again and again. It's simple: find three jewels in a row, then three more, then three more. iPad users should check out the app's big brother, Bejeweled Blitz, for the full HD experience.

Action and racing games

Temple Run 2

📱 Free (+in-app purchase)

Picture an endless Indiana Jones-style chase sequence and you'll get the idea. Run, slide and jump to make your getaway while collecting coins and power-ups. The first edition of this app was an instant classic and this sequel is even better.

Drag Racing Free

📱 Free

A well-respected racing game, with integrated Game Center duelling. Head into **Settings** and turn on vibrations to add a new dimension to the gameplay.

Nano Rally (HD)

📱 Lite/Paid

This fun little racing game sees your tiny vehicle going head to head with other minuscule motors. If you ever played the classic Micro Machines, with its wacky miniature driving environments, you'll love this.

TIP The best place to go online for the latest iPhone games reviews and news is toucharcade.com. The site's forums are very useful for picking up hints and cheats.

Rough Roads

📱 **Paid**

In Rough Guides' home-grown racing game, you drive your trusty camper van through a variety of global locations, collecting snippets of travel info as you go.

Atari's Greatest Hits

📱 **Paid**

This app gives you access to a large library of old Atari games from both the arcade and console vaults.

Angry Birds Star Wars (HD)

📱 📱 **Paid (+in-app purchase)**

More than just being the ultimate money-spinning franchise collision, this is a fabulous game and visually stunning.

Star Wars Arcade: Falcon Gunner

📱 **Paid**

Thanks to the iPad's built-in accelerometer you can spin 360 degrees to find and blast your targets, meaning that this game is best played on a swivel chair with pets well out of the way.

TIP If you've got an Apple TV and a recent iPhone or iPad, try using screen mirroring and AirPlay to play your games through a connected HD TV screen or projector.

Sports

FIFA 13

📱 Paid (+in-app purchase)

It's hard to fathom how EA Sports managed to get so much detail and sophisticated gameplay into an iOS game.

Flick Kick Football

📱 Lite/Paid (+in-app purchase)

If you're looking for a more retro soccer vibe, get your fingers limbered up for this onion-bag-filler.

World Cup Table Tennis (HD)

📱 📱 Lite/Paid

Highly addictive, this ping pong app really sucks you in. It hooks up with Game Center so that you can battle with opponents from around the world to win the championship.

Apple Game Center

Game Center is a social-networking tool that many app developers build into their games to enable global leaderboards and multiplayer gameplay. There is also an auto-match feature that can help you find new people to play with. The Game Center app comes pre-loaded on your iPhone and iPad; it gives you access to all your scoreboards and provides shortcuts to the actual game apps (tap through and hit the Play button). Find out more at apple.com/game-center.

8

Health & Fitness

Apps for your abs

Though self-diagnosis can be a dangerous thing (and should be avoided where serious symptoms are concerned), there are plenty of very useful and ingenious apps in the App Store that can help you take control of your personal health and fitness. This chapter homes in on a few of the best, in terms of ease of use and clever functionality, but this really is the tip of a very large iceberg, so do browse the rest of the category in the Store if you don't find what you need here. Also check out our recommendations for what's available in the Medical category (see p.71).

Measuring your health

The online Apple Store now has a dedicated section for Health and Fitness hardware to go with your iOS devices. It's mostly focused on the running and fitness markets, with the Nike-branded FuelBand gadget having received the most media attention. You wear it on your wrist and it monitors your workouts and daily activity, reporting back on your progress, and setting you targets, via an app.

WiScale

📱 Free

At the more specialized end of the market are accessories such as the Withings Smart Blood Pressure Monitor (pictured), which connects with this app to help you monitor your blood pressure. The same company also produce the bathroom

Wi-Fi Body Scale, which uses the app to monitor weight, fat mass and lean mass.

Moves

📱 Free

Leave this app running in the background and it will monitor how many steps you take each day (walking and running) and display the results on a geo-located daily timeline.

NHS BMI healthy weight

📱 Free

Of all the BMI apps tested, this is by far the most convenient and straightforward to use. It can remind you to update your weight regularly, and also gives you options to switch between multiple users and between metric and imperial measurements, which many of the competitor apps do not.

Calorie Counter (HD)

📱 📱 Free

There are hundreds of apps to track what's going in and what's being burnt through exercise. This has an easy-to-use interface and a massive database of food types and their calorific values.

Hydrate

📱 Paid

And for keeping track of your daily water intake, give this app a go. It's easy to use, notifies you when you forget to drink, and is less fiddly to use than some free alternatives.

Pillboxie

📱 Free

Though you could simply use alarms or a to-do list to remind you to take your meds, this little app does it in a very stylish, customizable way.

Instant Heart Rate

📱 Lite/Paid

An ingenious app that uses the torch and camera of the iPhone 4 and 4S (it also works with older models, but only in good light) to analyse changes in colour at the tip of your finger and then convert the data to a pulse reading. You can keep track of your heart rate before and after exercise, and build up a timeline of the data.

Sleep Cycle alarm clock

📱 Paid

With this app installed and your iPhone tucked under your pillow, you can monitor your periods of deep and shallow sleep (the app uses your phone's built-in accelerometer to measure when you're moving). Then, set the in-built alarm to only wake you at the most natural point in your sleep cycle. It works a treat.

Sleep Talk

📱 Paid

Ever wondered what you say when fast asleep? This app detects your midnight natter and organizes it into clips.

Condition

📱 Free (+in-app purchase)

This medical-log app gives you daily prompts to record your blood pressure, pulse, weight and temperature.

Personal trainers

Situps Coach

 Lite/Paid

Your fast-track to abs of steel, with detailed training schedules, calorie-burning stats and multi-user settings.

Ab Workouts (Pro)

📱 **Lite/Paid**

More tummy training, this time with twenty different exercises explained and illustrated. The same developers also offer Push Ups Pro and Butt Workouts Pro.

Headspace (on-the-go)

📱 **Free**

Train yourself to be calm and serene with daily guided meditations. The app also pushes "mindfulness buzzer" notifications to your phone to remind you to take a moment to reflect and relax.

WalkJogRun

📱 **Paid (+in-app purchase)**

This app offers interesting exercise routes based on your location (millions are mapped around the world) and training plans suitable for any fitness level, from beginner to marathon runner (some via in-app purchases), plus calorie-burn data.

RunKeeper

📱 **Free (+in-app purchase)**

Track your pace and running routes on integrated maps using the iPhone's built-in GPS chip.

Cycle Watch

📱 **Paid**

This cycle tracking and mapping app monitors your GPS position, elevation and speed, and keeps records so you can improve your times across the same route.

Pregnancy & babies

Pregnancy Day by Day

📱 **Paid**

Monitor your pregnancy with this app's insightful daily commentary, organized as an easy-to-follow timeline. Also includes a full search function, checklists, ultrasound pictures and reminders.

Withings WithBaby

📱 **Free**

This is another app that requires a separate piece of hardware to operate, in this case, the Withings Smart Baby Monitor. It even lets you watch your baby sleep in night-vision mode.

9

Lifestyle

Everyday apps & web services

Just like the Entertainment category of the App Store, the Lifestyle section is a diverse mix of the genuinely useful and the completely useless. Expect to find a lot of big brands offering alternative ways to reach their website content, alongside more functional apps dealing with shopping and, well, pretty much anything else you can think of. In the following pages, we try to make sense of shopping from your iOS device and also pick on a few other favourites from this section of the store to provide you with some good jumping-off points.

Shopping

eBay & eBay for iPad

📱 📱 Free

The eBay apps give you access to all the auction site's buying and selling tools, and instantly alert you if you've been outbid on an item. If you sell on eBay regularly, you'll probably find the app easier to use than the regular desktop site.

Fat Fingers

📱 Free

The Fat Fingers app searches eBay for commonly misspelled items to help you home in on the bargains that literate searches would miss out on.

Gumtree

📱 Free

The increasingly popular localized classified ads service has an intuitive app that lets you monitor your favourite listings and contact sellers directly if you want to buy their wares.

Etsy

📱 Free

Etsy is a retail portal (and now app) purely for unique craft and vintage items – essential for buyers and sellers alike.

Amazon Mobile

Free

A well-managed app for shopping, reading reviews and making purchases from Amazon. iPad users should also download the more visual Amazon Windowshop app.

Fancy

Free

Whatever takes your fancy (whether designer clothes or gadgets) can be found and added to your "fancy'd" feed or to a basket that allows you to make a purchase direct from the retailer.

Groupon

Free

Groupon is the big name when it comes to discount codes in the app world, though its selection can feel a bit random, so shop around some of the other voucher vendors too. Others worth looking at include Vouchercloud and Yowza.

Ocado

Free

Shopping for groceries online has been big business for a while now, so it was only a matter of time before the apps started appearing. Of all the UK supermarkets' apps, Ocado's offers the best user interface. In the US, search to see what's available in your area.

Miscellaneous

The DailyHoroscope

 Free

The most stylish of many horoscope apps in the store that deliver you a daily dose of astrological guidance.

Notabli

 Free

This cute app is for journaling moments in your kids' lives (photos, videos and text) and sharing the stories with family.

Rightmove

 Free

Finding property to rent or buy recently became infinitely easier thanks to the filtering tools of apps such as this (UK only). In the US, head straight to the ZipRealty app.

Blippar

 Free

Blippar brings cutting edge augmented reality (AR) experiences to cereal boxes, posters, book covers and the like, all via your iPad or iPhone's screen. Download it and have a play.

10
Medical
Anatomy, first aid and more

The medical industry is another arena that is rapidly being transformed by the world of mobile applications, smartphones and tablets. Medical students have access to extraordinary 3D and interactive tools that help them to understand the human body in a way that was previously only possible at the dissection table, while professionals have all sorts of tools at their fingertips to aid both diagnosis and patient monitoring. As for the layman, self-diagnosis (within reason) is easier thanks to medical apps, and everyone should have at least one first-aid app on their phone ... though hopefully you'll never have the opportunity to use it.

Anatomy apps

There are hundreds of different takes on the anatomy theme in the App Store, some aimed at kids, some at professionals, some at students. Many are just plain terrible in every way, while others are quite attractive, but shockingly inaccurate. Here are a selection that are worth taking a closer look at.

DK Human Body

📱 **Paid**

Picked by Apple as one of their "best of 2011" apps, DK's anatomy software lets you explore the various systems of the human body, rotating the on-screen subject as you go. Along with detailed, labelled diagrams and video content (complete with vibrations as you hear the heart beat and feel the nerve impulses race), there is also a useful self-testing mode aimed at medical and biology students of all ages.

iMuscle (NOVA Series)

📱 📱 **Paid**

The NOVA series of anatomy apps combine to produce a complete and indispensable guide to human anatomy. Though primarily aimed at the fitness market (this app illustrates various workout programmes and the muscle systems affected), iMuscle gives a good introduction to what the NOVA series is about in a very practical context. Other apps in the series include Muscle System Pro and Skeleton System Pro, both of which are more academic in tone and offer tools for annotation, 3D anatomy models and quizzes for self-testing.

iRis Interactive

📲 Paid

This virtual eye app recreates the functions of the eye, and the muscles around it, to help you understand how it all works and to create simulated eye-examination sessions. For an alternative eye app with a first-rate pedigree, check out EyeDecide HD.

Anatomy In Motion – Complete

📲 Paid

Not cheap, this very well constructed app uses inter-active animations to show you exactly what is going on in your body when you move. Bookmarking, notetaking and flashcard features all add to the value of the package.

Diagnosis

Just as with the internet, it's important to take all medical information found in apps with a pinch of salt. If you are in any doubt about your condition, or if your symptoms persist or worsen, you should seek advice from a medical professional as soon as possible, and not simply head back to the Store for another app injection. Still, you might find one of these to be a useful starting point.

Eye Chart

📱 Lite/Paid

This very well-designed app is comprised of various randomized eye tests which are designed to work on the iPad's screen and feel comfortable for the patient to view.

Hearing-Check

📱 Free

Specifically aimed at detecting sensorineural hearing loss (caused by noise or old age), and not damage to the middle or outer ear, this app walks you through a simple hearing test and then offers advice as to whether you need to see a medical professional.

Prognosis: Your Diagnosis

📱 Free

Aimed squarely at medical students, this fun diagnosis game lets you scrutinize clinical cases to see if you can figure out a patient's prognosis from the symptoms and information presented.

First aid

You never know when a first-aid app is going to come in handy, and it really could save someone's life during those crucial few minutes before the professionals arrive.

St John's Ambulance First Aid

📱 Free

One of the best apps in the Store for emergency situation diagnosis. The treatments and procedures are nicely organized and illustrated, making them both easy to find and easy to understand under what could be very stressful conditions. It's worth noting, however, that the procedures covered are based on UK protocols and so might contradict medical advice given elsewhere.

Army First Aid

📱 Paid

Based on a US Army field manual, this app is well illustrated, though at times a little wordy. The search feature is excellent, and reveals the depth of the content, covering everything from cuts and bruises right through to physiological trauma and biological weapon injuries.

First Aid

📱 Paid

Another good little app that covers basic illnesses and both accident emergencies and accident prevention.

Miscellaneous

Dragon Medical Mobile Recorder

 Free

One for the techie medical professionals out there, this app lets you dictate medical notes on the move and is fully integrated with eScription and iChart services.

AirStrip Cardiology

 Free

This award-winning, FDA-approved cardiac patient monitoring tool plugs into an existing AirStrip Cardiology medical environment to give touch-screen access to patients' histories. The same company also make a complete remote patient monitoring app called AirStrip – PM.

Patient.co.uk

 Free

This UK website now has a very handy app for browsing through hundreds of medical leaflets covering all sorts of conditions and diseases, and for finding various medical services in your area (England only at the time of writing). An alternative UK app that provides the latter service is Find GPs, while in the US, similar functionality (along with appointment managing tools) is delivered by the FindADoctor.com app.

11
Music

For fans, for musicians

For many people, the built-in Music app on the iPhone, iPod touch and iPad is as far as they take a relationship with music on their iOS device. But there's plenty more out there, whether you want to play music, stream music or make music. It's definitely worth exploring some of the other audio-related tools that litter the Music category of the App Store; in this chapter we pick out a few of the best.

Apps for listening to music

Audium

🔲 Paid

This gorgeous, award-winning alternative music player for the iPhone or iPod touch focuses on album (rather than playlist) playback, through some very smart gesture-based controls.

Spotify

🔲 Free (+subscription)

With a subscription, you can stream unlimited music to this app from the internet (and also play downloaded selections offline) for around the price of a CD album per month. Without the premium account, you can stream music to the app from your desktop (a bit like Home Sharing). Though not yet available in every country, it's pretty compelling for those who can get it.

> **TIP** Also download Remoteless (for Spotify) – the best remote control app for the desktop version of Spotify.

AccuRadio

🔲 Free

There are plenty of internet radio players to be found in the App Store; this one is free, has an easy-to-use interface and lets you create favourites lists.

TuneIn Radio

📱 Lite/Paid

Another great radio tuner app; you really get your money's worth with the Pro version, including a sleep timer, pause and rewind controls, and a filter tool that chooses stations for you based on your Music library content.

7Digital+

📱 Free

As an alternative music download service to iTunes, 7Digital (via a desktop machine) is a good choice. Once set up, all your music will sync over Wi-Fi to this app for playback on the move.

Couch Music Player

📱 Paid

A nice alternative to the iPad's built-in Music app, this piece of kit features drag and drop playlist queue creation, full-screen artwork mode and some nice gesture controls.

Bloom.fm

📱 Free (+subscription)

With a very unusual flower-styled interface, this app puts a novel spin on the music-library model. You get to "borrow" a certain number of tracks from the Bloom cloud and keep them locally, offline, trading them in when you want something new.

VinylLove

Paid

For the nostalgic among you, this virtual turntable lets you listen to the music from your regular Music app library on your iPhone or iPad with a little added crackle. You also get to move the turntable's arm to choose a track or a position within a track. For a retro cassette vibe, download the AirCassette app.

Deej

Paid

With this app you can choose tracks from the library on your device and then mix things up, DJ-style. With tools for scratching, gain, pitch, crossfade, bass and treble, the experience is pretty authentic. You can also save cue points and record your sessions.

AutoValve

Paid

With its sophisticated mixing and filtering tools, this app lets you add a retro vibe to the songs playing through your Music app. Choose from vinyl, valve, reel-to-reel and 8-track filters, and watch the VU meters twitch.

Virtual instruments

The App Store's Music category is awash with virtual instruments, sequencers, drum pads and other noisy creations that, in the hands of most people, would make you want to stuff cotton wool in your ears. With a bit of perseverance and a sprinkling of talent, however, it's possible for some of the apps to become more than just a novelty.

GuitarToolKit

 Paid

This is one of the best virtual guitar apps. It comes complete with a tuner and metronome to help you play one of those old-skool wooden versions.

Seline Ultimate

Paid (+in-app purchase)

Both a unique ergonomic playing environment and an amazing selection of synthesized instruments, filters and audio tweaking tools. The app is also fully MIDI compatible, meaning you can hook your iPad up to a keyboard and play using conventional keys. Though there's more than enough to keep you busy, the in-app purchase library includes extra loops, drones and tones.

TNR-i

Paid

Innovative electronic instrument that uses a grid of glowing buttons to create futuristic music.

Piano/Piano Tunes

 Free/Paid

Nice-sounding (and many aren't) piano app with tunes to learn. Piano is free, while Piano Tunes is paid and ad-free.

NLog MIDI Synth

 Paid

Arguably the best Korg-style tone generator for the iPhone and iPad, with a delightful analogue feel to the interface.

DXi FM synthesizer

Paid

Inspired by the classic 80s DX7 synth, this app features a great sound, nice interface and also supports Core Midi connectivity.

Soundrop

Free

Draw and drag lines and shapes to make music. It's easy to pick up and very addictive once you get a feel for how the geometric parameters influence the tones.

TIP Remember that the headphone jack can be used as a line-out to play your creations via a PA, amplifier or hi-fi system instead of your iPad or iPhone's built-in speaker.

Studio solutions

BeatMaker 2

 Paid

Before testing this app on the iPhone you might not think that serious music creation was possible on the smaller screened iOS devices … but what you can achieve here is exceptional. Featuring a drum machine, sequencer, sampling keyboard, mixer console, wave editor and more, this app is an undisputed bargain. The iPad experience is even better.

GarageBand

 Paid

Apple's all-in-one music creation package comes into its own on the iPad. There are both virtual instruments and so-called "smart" instruments (which pretty much play themselves) and a very swish multitrack mixing environment with a plethora of filters and adjustments. The app also allows compositions to be exported easily via the iTunes File Sharing area for use in GarageBand or Logic on a desktop machine.

FourTrack

 Free

This is an incredible multitracking mini recording studio that fits in your pocket. It can be a little fiddly to use, but it's also very useful for quickly getting song ideas down.

Looptastic Producer

 Paid

An amazing pocket dance-music creation tool, with loads of built-in loops and the ability to add your own. There's also a bunch of effects and time-stretch tools.

StudioTrack

Paid

Another fully featured recording environment for aspiring musicians and songwriters, with a very easy-to-use (and pleasing on the eye) interface.

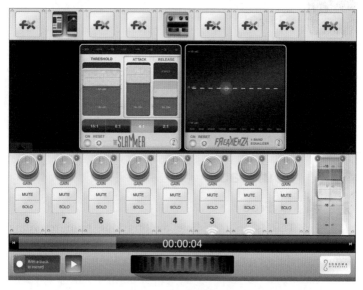

NanoStudio

 Paid

Before making a final decision on which iOS recording studio is going to suit you, check out this contender. The added ability to upload straight to SoundCloud is a useful bonus.

Miscellaneous

Songsterr Guitar Tabs

Paid

More guitar and drum kit tabs laid out in one place than you can shake a stick at. This also features a built-in guitar engine, so you can hear what the tab should sound like before you have a go, and a control for adjusting the tab scroll speed.

Songify

Free (+in-app purchase)

This has been a real online sensation. You basically sing a few lines into your phone and the app instantaneously transforms it into a tuneful, groovy pop hit.

TIP The iPhone comes with a built-in audio recorder called Voice Memos, which automatically transfers recordings you make back to iTunes when you sync. On the iPad, try an alternative such as Audio Memos.

Nota & Nota For iPad

Paid

A fantastic music theory app, featuring a piano chord and scale browser, a landscape keyboard for practising and a quiz for testing your progress.

Discovr Music

Free

An essential download for all music lovers, this app shows you all the bands that it thinks you might like based on your favourite artists and genre choices. You can watch the bands' videos and hear song previews – it's a great way to find new music.

SoundCloud

Free

The SoundCloud site and community is a great place to both discover new music and post your own creations.

SoundHound

Lite/Paid

There are many music recognition apps out there that can hear any song playing and tell you what it is. Shazam is arguably the best known, but SoundHound is our recommendation. Aside from being fast and reliable, it even does a pretty good job of recognizing hummed tunes as well as the recorded originals.

12

Navigation
The best map & location tools

The App Store has separate categories dedicated to Navigation and Travel, in both of which you'll find everything from plane and ship location tools to multifunctional compass apps. And because the category choices are at the discretion of the individual developers publishing each app, there don't seem to be any golden rules along the lines of "map apps are always in Navigation" or "hotel apps are always in Travel" to guide you. So, have a browse to see what takes your fancy, and if you can't find what you're looking for here, try the Travel chapter too (see p.155).

Maps

Search "maps" in the App Store to unearth some interesting cartographic relics (such as Historic Earth), as well as some very impressive reference tools (such as National Geographic World Atlas). You'll also come across…

Google Maps

📱 Free

The introduction of Apple's own mapping system in 2012, as the default in place of Google's, was not their finest hour. The media had a field day identifying the mapping inaccuracies and many users refused to install the software update. Several months later Google launched their own app for the iPhone. It's an essential piece of kit with well-rendered mapping and excellent routing tools.

Outdoors GPS Great Britain

📱 Paid (+in-app purchase)

Reliable UK Landranger mapping, courtesy of Ordnance Survey. Use the in-app purchases to download UK regions, and though they're not cheap, you do get a lot of coverage in each one.

Map+

📱 Paid

Create your own custom maps using the built-in set of colourful and varied pins and icons. You can also usefully attach notes, addresses and photos to individual points of interest.

Google Earth

Free

An essential free app that gives you unfettered access to the globe. When you swipe with two fingers, or alter the tilt of your phone or iPad, you adjust the pitch of your view, which allows you to achieve some amazing vistas of mountainous regions. Some cities (San Francisco, Boston and Rome, for example) are fully rendered in 3D for the ultimate fly-through experience.

Transit Maps

Free

Use this app to download and store transit maps from around the world to use offline. It's also worth looking into alternative city-specific transit apps, as many will give you the latest service information, assuming you have a web connection (worth checking before you head down the stairs to the underground station).

Tube Exits

Paid

For the London Underground, this is *the* essential app, with network updates, a route finder and an offline map. Best of all, however, is the app's key selling point: the fact that it tells you which end of the train to board to be near your exit when you get out at the other end. The same company also publish the Paris Metro Exits and Berlin U-Bahn Exits iPhone apps. If you are after something similar for the iPad, then take a look at the Station Master – London Tube Edition app.

ViewRanger Outdoors GPS

 **Paid (+in-app purchase)**

This is one of the best apps around for cyclists, walkers, climbers or other outdoor types. The in-app purchase mechanism gives you access to offline mapping from a variety of sources, including Ordnance Survey, OpenStreetMap and OpenCycleMap, and the built-in tools for planning and logging routes are excellent.

Maplets

 Paid

A great collection of offline national park and bike trail maps for both the US and international destinations (not currently available in the UK App Store). There are also hundreds of ski maps and transport maps available to download and use later when you don't have internet access. And when you are online, you can also pull down general weather and snow reports.

Greater London A-Z

 Paid

Brilliant sets of London street maps (there are several editions for both iPad and iPhone to choose from, depending on your need). All the mapping is stored offline, so you don't even need a web connection to use it. Equivalent offline maps are available in the App Store for most major cities – essential if you're going abroad and don't want to rack up expensive roaming charges by using a mapping system that requires a data connection.

GPS Navigation

If you have a recent iPhone or iPad, and you are running the latest software, then turn-by-turn navigation is at your fingertips as part of the built-in Maps app's toolkit. There are, however, alternatives:

TomTom

📱 Paid

These GPS navigation apps might be pricey, but they're still cheaper than buying the equivalent TomTom hardware ... though not that much cheaper. Beyond the expected turn-by-turn navigation, you'll also find regular map updates, speed camera alerts and integration with your Contacts app.

GPS Navigation 2

📱 Paid (+in-app purchase)

Buy the OpenStreetMap-based app once and then download individual data sets for your area (Europe, UK, etc) at very reasonable prices. Though they don't yet carry the same heavyweight brand recognition as TomTom, it's hard to see how Skobbler (the company behind this app) can go wrong with this approach.

Social GPS

📱 Paid

This is one of the many emerging services (including high-profile ones from the likes of Apple, Google and Facebook) that let you share your location so that friends can find you.

Miscellaneous

Milebug

 Lite/Paid

This very handy mileage tracker is an essential piece of kit for anyone that submits their own tax returns and needs to record and calculate travel expenditure deductions.

Spyglass – AR Compass

 Paid

A fully-featured, augmented reality toolkit that's a great alternative to the iPhone's built-in Compass app; and it works well on the iPad too. The integrated Gyrocompass is particularly impressive.

Ski Tracks – GPS Track Recorder

 Lite/Paid

Record and map your days on the slopes with Google Earth 3D mapping integration, altitude and stats reports, and pin geo-tagged photos to your routes. Best of all, you can do all this with data roaming switched off.

Current Elevation

 Free

Potentially handy little app that tells you both your current altitude and also the "ground elevation" of your location.

13

News & Newsstand

Keeping in touch with news and views wherever you are

Both the iPhone and iPad are ideally suited to delivering your daily fix of current affairs, or any other kind of news for that matter, whenever and wherever you want it, and in a perfectly digestible format. So, whether you read in bed, on the train, on the couch or at the breakfast table, there are apps that can act as a handy replacement for any newspaper, magazine or periodical.

News apps

Many of the major news and magazine publishers are battling to define their place in the digital marketplace as a whole. Some media companies have opted to make their apps available for free, but with advertising support; some are trying to get users to pay a subscription; others charge a one-time fee for users to download their app.

The real question here is not whether you want to pay for your news (there are still a thousand places online where the latest stories can be harvested for free), but whether you are prepared to pay for a particular stance, attitude or editorial voice – which is what we do in the real world when we choose to hand over cash for a particular print newspaper. Here are a few recommendations:

BBC News

📱 Free

This app has a well laid-out interface in both landscape and portrait modes, complete with a handy news-ticker that dishes up the latest headlines and embedded video content alongside the written news stories. You can customize the news categories that you want to see and share stories vie email, Facebook and Twitter.

The Guardian

📱 Free (+in-app purchase)

Great-looking subscription UK news app for the iPhone with loads of photography and a customizable download feature, so you only get the news categories you want. To read *The Guardian* on the iPad, look within the Newsstand category (see p.97) for a separate version of the app.

WSJ Live

🔲 Free

To read the regular *Wall Street Journal* app head to Newsstand (see p.97) and subscribe from there. This app dishes up video news clips from WSJ correspondents around the globe – great for breaking news and business analysis.

USA Today

🔲 🔲 Free

A really nice, completely redesigned interface with an easy-to-navigate sidebar control. Scrolling from page to page in long articles is particularly well handled, and the embedded video clips are an added bonus. Technology and US news are covered particularly well, while the localized weather forcast is an added bonus.

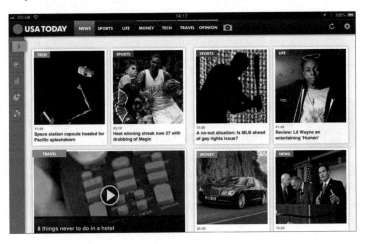

Longform

 Paid

An impressive magazine-style curated reading experience, Longform is worth taking for a spin if you want to get a cross-section of articles and news from around the world, but without having to do all the hard work of subscribing to various newsfeeds and the like.

Newsy

 Free

Award-winning video news app designed for snacking on stories when you haven't got time for the heavyweights.

TIP The Financial Times no longer has a dedicated app in the App Store; instead you need to use their excellent web app (app.ft.com). It's free to browse a small amount of teaser content, but you have to sign up for a subscription to get the full package.

Zinio

Free (+in-app purchase)

This app is a one-stop shop for magazines. There are hundreds on offer and plenty of free samples, and you can choose to purchase either single issues or annual subscriptions. There's also a very handy calendar view that shows you which issues you'll be getting and when.

Newsstand

 **Built-in**

Tap the built-in Newsstand app icon on your Home Screen to get quick access to subscription magazine and news services that publish into the Newsstand category of the App Store, but without the hustle and bustle of the regular News category. To furnish your Newsstand with publications, either browse via the App Store or tap the shortcut button at the top of the expanded Newsstand panel. In most cases, you can then subscribe or purchase new issues from within each publication.

Prismatic

 Free

Connect this app to your Facebook, Twitter or Google+ account to receive news content based on your interests (which are inferred from your social media activity).

Currents

 Free

Google's curated news app dishes up content from most of the major news organizations as well as many book publishers. The content is broad in subject and there are both video clips and prose articles to be had. The app also integrates Google Translate functionality meaning you can read content produced in up to 44 languages in your native tongue.

News aggregators

Many news apps use RSS – Really Simple Syndication – to pull "feeds" or "newsfeeds" from blogs, news services and other websites. Each feed consists of headlines and summaries of new or updated articles. If you see something you think you'd like to read, click on the headline to view the full story. You can use a tool called an aggregator or feed reader to combine the feeds from all your favourite sites. It's almost like having your own personalized magazine or newspaper.

At the time of writing, arguably the best web-based aggregator service is Google Reader, which can be set up via reader.google.com, using the same credentials you use for other Google services.

As for dedicated apps for reading your aggregated newsfeeds, try one of these:

Feeddler RSS Reader Pro

📱 **Lite/Paid**

There are free versions of Feeddler available, but they don't offer anywhere like the same toolkit. If you want Evernote and Instapaper integration, as well as custom sync options and automatic URL-shortening when you share stories, go straight for this Pro version.

Early Edition 2

📱 **Paid**

This iPad-only aggregator creates a dynamic custom layout based on the feeds you point it at. It feels like a newspaper to read, and has some very nice extras, such as a gallery mode that pulls together all the images associated with your stories.

Pulse

 ### Free (+in-app purchase)

One of the more beautiful news aggregator apps, Pulse pulls together newsfeeds and social media feeds from multiple sites to create a mosaic of content that is both easy and fun to digest. In exchange for an in-app purchase you can sign up for additional premium content.

Newsify

 ### Paid

This elegant aggregator integrates cleanly with your Google Reader account and offers a plethora of features, including an offline mode (with image syncing), a night-reading mode and fully editable subscription lists.

Zite

Free

This curated collection of news stories unearths some pretty interesting stuff that you might not catch in other nets.

Flipboard

Paid

The very popular Flipboard pulls together news, website feeds and social media feeds (from Facebook, Twitter and others) to create a very stylish, personalized magazine-like experience.

Reading webpages later

Also worth investigating is the Reading List feature built into Safari on the iPhone and iPad. When you find a webpage but can't read it right away, tap the ⬆ button and choose "Add to Reading List" to save it for later. You'll find the Reading List at the top of your Bookmarks menu, accessed via the ⬚ button in Safari.

> **TIP** If enabled within **Settings** > **iCloud**, your Reading List will sync with your other Apple devices along with your Bookmarks.

Instapaper

 Paid

If you don't get on with Apple's Reading List, sign up for the excellent Instapaper service (instapaper.com). The app has an easy-on-the-eye, stripped-back feel; many RSS services integrate with it, as does Twitter, so you can keep all your reading matter and stories with you, and it's also available offline (assuming you remember to sync your app and Instapaper account before you disconnect from the internet).

Pocket

 Paid

This excellent bookmarking app has a really clean interface. Once you find things that you want to come back to later, add them to Pocket and they are synced across devices and desktop machines where you have your Pocket account set up.

14

Photo & Video

Making the most of your camera

There's a lot you can do with the cameras found on the iPad and iPhone straight out of the box, but there's also plenty more to be discovered. Whether you want to craft near-professional quality photos or re-create the look and feel of a bygone age of film-making, there are hundreds of apps out there to help you. Here are a few of our favourites.

Camera apps

Camera+

 Paid

This is an excellent, fully featured Camera app. Its special image stabilizer control works really well to stop the wobble.

CameraBag 2 (HD)

 **Paid (+in-app purchase)**

Claiming to be the "first hi-fi camera app", this toolkit boasts a very nice collection of filters and visual effects to enhance your iPhone and iPad photography. The in-app purchase gets you a bunch of extra styles along with tools to create custom styles.

Hipstamatic

 Paid

Re-live those analogue days of oversaturated, grainy snaps with this popular app. You can share via Facebook and Twitter, and there are tools for organizing your images into "stacks".

KitCam

Paid (+in-app purchase)

With white balance, exposure control, lenses and various different film stocks, this is a rich alternative to the iPhone's Camera app for shooting both stills and video.

Instagram

📱 Free

Though there was controversy at the tail end of 2012 around the privacy policy of this Facebook-owned photo-sharing toolkit, the company moved swiftly to sort out the situation, so this very popular app is most definitely worth installing.

Wood Camera

📱 Paid

Another amazing armoury of camera lenses, filters and tools, and a nifty light-box feature.

FatBooth

📱 Paid (+in-app purchase)

This little app shows you what you'd look like if you were more interested in pies than iPhones. There's also an in-app purchase to create overweight video clips as well as stills. The same developers also make MixBooth, which melds two faces to reveal the potential love child of any number of unlikely combinations

Group Shot

📱 Paid

Snap several shots of a group of people, then swap heads between the images. It's a bit of a cheat, but it does mean you don't end up with that one person with their eyes closed.

AutoStitch

 Paid

If you own an iPhone 5 or recent iPad then you will be able to capture panoramas using the built-in Camera app (tap Options to get started). As an alternative, this app can be used to create stitched panoramas from multiple shots taken from either your iPhone camera or your existing photo library.

360 Panorama

 Paid

This panorama app is easy to use and is backed by a growing online community. Your panoramas are uploaded to the cloud where the images get automatically re-tweaked and processed, giving you an even better final version of the shot.

Photosynth

 Free

Another impressive 360° panorama-capturing app, this time from Microsoft. One-shot capture in all directions means you can theoretically photograph an entire sphere.

Vapp

Free

This app lets you … should you need to … trigger your iPhone's camera shutter using a pre-programmed sound.

Image editing and adjustments

Photos

 Built-in

It's worth checking out the adjustment tools built into the Photos app. Select an image and tap **Edit** at the top of the screen.

Adobe Photoshop Touch

Paid

Anyone familiar with the professional desktop application is going to feel very at home here, with layers, adjustments, filters and standard selection tools all on offer.

Tiltshift Generator

Lite/Paid

This app has a toolkit of filters and blurs that re-create some of the hallmark effects of retro toy cameras, which used cheap lenses and were largely made of plastic. You can also use this app to create interesting focus effects within existing images.

Photogene

Paid

The built-in editing tools of the Photos app don't come close to matching what's on offer here. Dive straight in and adjust curves and colour graphs, and add art effects and speech bubbles.

Snapseed

 Free

Another top-class app for adjusting photos and images to create some very professional results.

iPhoto

Paid

With the arrival of the third-generation, Retina display iPad, Apple also unleashed iPhoto as an iOS app. Its toolkit of filters and adjustments is very impressive, and the whole thing is managed using some very intuitive and clever gesture controls.

Online image posting

Aside from Apple's own Photo Stream tool, there are hundreds of apps for photo sharing and uploading just waiting to be discovered in the App Store. Flickr, the world's most popular photo-sharing site, has an excellent free app that account-holders can use. If Picasa is your preferred online photo storage solution, then look out for an app called simply Up, and for Facebook, you really do have hundreds of apps to choose from to help with mass uploads. For the iPad, the best choice is PhotoLoader HD (icon pictured). Instagram is another decent choice: it features a wealth of filters and effects for styling your images as well as options for uploading to Facebook, Twitter, Flickr, Foursquare and others. It's also worth noting that Facebook have a dedicated Facebook Camera app (icon pictured).

Finally, if you've entered the appropriate account details within Settings > Twitter and Settings > Facebook on your iPhone or iPad, you can post to Twitter and Facebook directly from both the Camera Roll and albums within Photos. Simply tap the ↪ button and choose the Twitter or Facebook option.

Relievos

 Free

This app takes regular images and allows you to turn them into 3D "pop out" images. It works with some images better than others, and it's quite a tricky process to get the hang of, but once you've got the feel for it, the results can be very impressive.

Color Splash

 Paid

Isolate and then colour and adjust individual sections of your images to create some novel results.

General photography

DK Tom Ang's Digital Photography

📱 Paid

From professional photographer and writer Tom Ang, this is one of the best photographic guidance apps available. As well as a wealth of general tips for framing and composing your shots no matter what kind of camera you use, there's also a great filtering tool for choosing the information you need based on the subject matter and conditions – all conveniently available on your iPhone.

SLR Photography Guide

📱 Paid

This handy field guide dishes up short tutorial videos to help you get the most out of your digital SLR camera, along with articles on various camera settings and features. Though at times quite hard to navigate, it's sure to offer some good tips.

FotometerPro

📱 Paid

Though this app might not supply the precision that some photographers need, it offers both reflective and incident modes and a nice retro interface, so goes some way to providing a light-meter tool for your iPhone or iPad.

Video apps

iVideoCamera

 Lite/Paid

One for the early adopters, this video app is for pre-3GS iPhones that didn't support video recording straight out of the box.

iMovie

 Paid

A great app for adding sophisticated editing effects, transitions, themes, music and so on to your movies. Though the trailers feature and some of the themes might seem a little cheesy, the timeline and transition tools work well on the iPhone's small screen. As for the iPad, this app really shines on the third-generation Retina display models.

1 Second Everyday

Paid

This app is looking for a bit of commitment from you: the idea is that you record one second of video every day and then, after years of habitual use you end up with a dizzying movie of your life and times. You can have multiple timelines running simultaneously (one for each of your kids, perhaps) and you can also set up reminders to nudge you when you might have forgotten to shoot. The app backs up all your content to iCloud, so you might, theoretically, still be working on the same video in a couple of decades' time when the iPhone 25S comes out.

iSuper8

Paid

A really nice little app for shooting Super 8-style footage. Unlike the original film format, you get the bonus of audio thrown in, and you can adjust various settings, including the frame rate and the level of "grain".

8mm Vintage Camera

Paid

Another good option for the vintage 8mm feel. This app features some nice gesture controls to add flickers and jitters.

Cinemagram

Free

This lovely piece of software gives you the toolkit to create intriguing hybrid movie/photo clips with vintage filters liberally sprinkled over the top. You can then share your creations through Twitter, Facebook or Tumblr.

iMotion HD

Free (+in-app purchase)

This stop-motion toolkit also features time-lapse functionality. Download the separate iMotion Remote app to preview and capture frames remotely. The option to export your creations is available as an in-app purchase.

iStopMotion for iPad

🔲 Paid

All you need to film your own stop-motion animations at home. The app shows you the previous frame as you reposition your subject to create an impressively smooth effect. The same company also offers the free iStopMotion Remote Camera app, which lets you take snaps using a Wi-Fi-connected iPhone (or another iPad) while previewing the animation on an iPad running the full paid app. It's an awful lot of fun, and intuitive enough for both adults and kids to get to grips with.

Frameographer

🔲 Paid

Create incredible time-lapse and stop-motion movies with this app (in HD if your iPad or iPhone model supports it). For time-lapse projects you can set the gap between shots anywhere up to 24 hours apart, and when you're finished you can add your own audio soundtrack or narration.

Speed Machine

🔲 Paid

This video-shooting app lets you adjust the capture rate of your iPhone or iPad so you can create both slow motion and speeded-up movie clips.

Long Exposure Photography

Paid

This app turns your iPhone into a long-exposure photography studio by processing video clips. It works best in low-light situations where you are trying to capture points of light ... car headlights, fireworks, and suchlike.

Backwards Cam (HD)

Paid

This easy-to-use tool for reversing video clips gives you speed options, soundtrack functions (choose music from your library or reverse the original audio) and trimming tools.

CineBeat – Music Video Maker

Free (+in-app purchase)

Shoot regular video clips and then add filters, transitions and special effects to create impressively professional-looking mini movies. The sharing tools make it very easy to pass on your creations, and the in-app purchase options give you loads of extra sparkle, themes and the ability to create longer clips.

15
Productivity
Tools for everything

Like most modern computers, the iPhone and iPad come with a small menagerie of extra tools built in. Of the two, the iPhone is better kitted out, featuring a calculator and weather app, while both devices come furnished with a fully spec-ed Clock app (alarms, timers, etc), the useful, but basic Notes app and the Reminders app (see overleaf). Thanks to the App Store, however, you can equip your device with tons of useful, feature-packed apps.

To-do lists

Reminders

📱 📲 **Built-in**

This app syncs with iCloud (turn it on within **Settings > iCloud**) and with iCal (called Calendar in Mountain Lion) and Outlook, and allows you to set reminders on the iPhone 4S using the Siri voice assistant. You can also create location-based reminders, so, for example, your phone may remind you to buy some milk when you enter the supermarket.

Remember The Milk

📱 **Free (+subscription)**

Speaking of buying milk ... this app is arguably the best task-organizing sync service available right now, though you do have to sign up for a fee-paying Pro account to make use of the iPhone and iPad apps. Like Reminders, it supports Siri.

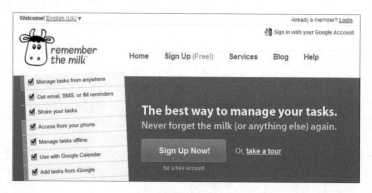

DOOO To-do note

📱 Paid

With a fresh and practical approach to to-do list management, this app lets you voice-record your tasks, add them as text and even draw them (great for getting ideas down swiftly). Overall, it feels as quick and easy to use as a stack of sticky notes.

GeeTasks

📱 Lite/Paid

A slick little app for Google Task syncing, complete with a Home Screen badge icon for uncompleted tasks.

Do5

📱 Free

With customizable fonts and wallpaper background images, this app lets you create a simple, but effective, five-item to-do list on your iPhone's Lock Screen.

OmniFocus

📱 📱 Paid

If you like your to-do list tools to be complicated … check in here. The app isn't exactly cheap, but there are no ongoing subscription charges, and with syncing (to other OmniFocus applications) and a ridiculous amount of customization and task tools, you do feel like you're getting your money's worth.

Note-taking apps

Notes

📱 Built-in

Notes is Apple's simple application for jotting down thoughts on the go. Tap ✚ to bring up a blank page and you're off.

Awesome Note

📱 📱 Lite/Paid

A really stylish notes and to-dos tool with colour-coded folders, lists, alerts, journals and support for Evernote and Google Docs account syncing.

Evernote

📱 Free (+in-app purchase)

Evernote is among the best note-taking tools and services available. You can create notes from text, audio and images (any text included in pictures is automatically recognized and, impressively, becomes searchable), and the notes are then synced between the Evernote server and your Evernote applications.

Catch Notes

📱 Free (+in-app purchase)

Another good note-taking option, with solid cloud backup and a decent set of collaboration tools.

Nebulous Notes

Paid

This elegant note-taking app syncs with Dropbox (see p.129), features loads of built-in editing tools and themes and also has a distraction-free mode similar to that of WriteRoom (see p.119).

Springpad

Free

This is a "save for later" app with a twist. Add voice notes, text notes, locations, photos and more, and then when you come back to them, Springboard has provided useful additions to your notes, such as directions, price comparisons and reservation links.

Audio Memos

Lite/Paid (+in-app purchase)

This is a great audio-recording app for both iPhone and iPad users. The sound quality is high, and there are many filtering, editing and sorting options. If you get really serious, look to the in-app purchases to add compression, voice activation and so on.

Audiotorium

Paid

Unlike many note-taking apps, this gives you a full set of rich-text functions (bold, italic, etc) and the ability to record hours of audio while also taking text notes, making it great for lectures.

Word processing & writing apps

 ## Pages

 Paid

Apple's word processing app is very slick, but pricey compared to others. It does integrate nicely with iCloud though, meaning you can pick up where you left off on another device.

 ## NewPad & NewPad Plus

 Lite/Paid

For a free word processing iPad app, the lite version of this app has some great features; the gesture controls for cursor control are particularly useful. The paid version adds in Dropbox syncing, justification, word count, HTML export, and more.

 ## WordPress

 Free

For WordPress bloggers, this app is a must: it's both well designed and easy to use. As well as creating and editing your posts, you can also moderate comments and add audio and video.

 ## Day One (Journal/Diary)

 Paid

This is, by far, the best diary-keeping app in the store, with loads of sync options and a lovely calendar view.

Celtx Script

 Paid

If writing screenplays and scripts is your thing, this app (and connected cloud-syncing service) should be right up your street.

WriteRoom

 Paid

Offering a distraction-free writing environment without toolbars or formatting worries, this app also has a nice white-on-black mode that's easy on the eye for extended periods of writing.

Pocket CV

 Paid

This app ensures that you always know where your résumé is. It also has impressive layout features, meaning you can email a fully fledged PDF CV at the drop of a hat, from anywhere.

Studio Basic

Lite/Paid

This rather fine notes and journaling app is designed to be used with Byzero's (pricey) stylus pen, which you'll need to buy separately. The pen has a very fine nib that – combined with the app's ability to ignore your hand resting on the iPad while you write – gives a really nice user experience. Check out by-zero.com or Amazon, for news and reviews of the hardware.

Number apps

Calculator

 Built-in

Paying no small tribute to the classic designs of German functionalist Dieter Rams, the look of the iPhone's Calculator app can be traced back to a calculator that Braun (for whom Rams worked) built for Apple in the 1980s. When your phone's in landscape position, the calculator automatically changes to scientific mode.

PCalc Lite & PCalc

 Lite/Paid

If you have an iPad, you'll need a calculator to run on it. PCalc Lite is free, looks nice, has scientific functions and can also handle unit conversions. Upgrade to the paid-for PCalc RPN for additional scientific functions and layout options.

Numbers

 Paid

This iWork application is Apple's answer to Excel. Though not as sophisticated as the Microsoft desktop program in terms of advanced spreadsheet features, it is nicely put together and does an impressive job of creating both 2D and 3D charts and graphs. The templates are excellent and you can move between dry-looking data and something eye-popping with little effort. The Share & Print function allows you to send files via the iTunes File Sharing mechanism as a PDF, Numbers or Excel document.

Llumino

 Paid

This elegant calculator app may not have the advanced features of our other recommendations, but does boast some nice glowing visual effects.

Quick Graph

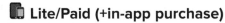 **Lite/Paid (+in-app purchase)**

A nice app that can be used to produce both 2D and 3D graphs from complex formulas and equations. The in-app purchase mechanism in the lite version gives you the advanced features of the paid app: implicit graphs, tracing, etc.

Sheet2

 Paid

This app crafts fairly impressive spreadsheets and can both edit and create Microsoft Excel documents. It also works with Google Docs for sharing your files online. If you like the sound of Sheet2, check out the Doc2 (for word processing) or the Office2 app, which combines the two at a cheaper price.

Analytics for iPad

 **Free (+in-app purchase)**

One of the best Google Analytics dashboard apps, with loads of settings as well as syncing and reporting tools.

Presentation apps

Keynote

 Paid

The third member of Apple's iWork toolkit is Keynote: a stylish equivalent of Microsoft's PowerPoint presentation software. What it lacks in features it makes up for in ease of use and elegant templates. It can open and edit PowerPoint documents, and you can also save Keynote projects to iCloud and then download them to a desktop machine as Keynote, PDF or PowerPoint documents.

> **TIP** Connect your iPhone or iPad to a projector using an iPhone Dock adapter or AirPlay, and you can use Keynote to run your presentation as well as to create it.

Keynote Remote

 Paid

This Apple app turns your iPhone or iPod touch into a wireless remote, letting you control a Keynote presentation running on your Mac or another iOS device over the same Wi-Fi network.

MyPoint PowerPoint Remote

 Lite/Paid

Similar to the Keynote Remote, this app is for controlling PowerPoint presentations playing on either a Mac or a PC. The lite version is iPhone-only.

File Sharing with apps via iTunes

Many apps, including the three members of the iWork suite (Pages, Numbers, Keynote), can utilize the File Sharing feature of iTunes to transfer files back and forth between your iPhone or iPad and your computer. If an app supports the feature, it will appear within the File Sharing panel in the lower area of the Apps tab in iTunes when your device is connected and highlighted in the iTunes sidebar. From there you can drag files in (or use the **Add** button to browse for files) and drag files out (or use the **Save to** button to browse for a location to save to).

If you can't see specific iWorks files in iTunes, you will first have to go back into the respective app on your device, head to the documents gallery, tap **Edit**, make a selection and then tap ↪ to get the option to move your document to the iTunes File Sharing panel.

Similarly, to pull documents into the iWorks apps, go to the app's document gallery, tap the ✚ icon and choose the iTunes option.

If you'd rather share files between your device and computer over Wi-Fi without iTunes, use an app such as Air Sharing (p.130), and for sharing via the web, try Dropbox (p.129) or iCloud.

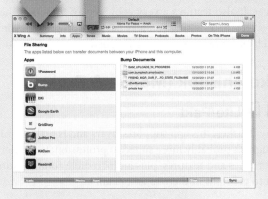

ProPrompter

📱 **Paid**

With this app you can turn your iPhone or iPad into a script teleprompter and, with the same app installed on another device (iPad, iPhone or iPod touch), you can control the pace of the script on the first device via a Bluetooth connection.

Artistic apps

Brushes 3

📱 **Free (+in-app purchase)**

Brushes is a beautifully crafted app that artists of any level can have a lot of fun with. It supports image layering, different brush textures and weights, and the ability to control the transparency and weight of your strokes based on the speed with which you move your finger. It also features a very cool "replay" function that records every brush stroke, letting you play back the progress of your creation to watch as a movie once you've finished.

SketchBook (Mobile & Pro)

 **Lite/Paid**

Like Brushes, this app supports layers and is really intuitive to use, though it's probably more suited to professionals. It offers loads of tools, brushes and textures, plus an incredible zoom feature that allows you to get right in there to add detail.

123 Color HD

📱 Paid

There are loads of colouring apps aimed specifically at kids. This one is a real treat – it has songs, voiceovers and multiple languages built in (so it can be used for basic language teaching too). When they're done, your kid's masterpieces can either be saved to Photos or emailed to granny and grandad.

SketchTime

📱 Paid

Specifically designed with realistic sketching in mind, this app dispenses with all the artist clutter and concentrates on the screen contact experience. The strokes and weights feel very natural and there is also a mode for tracing imported photos.

Whiteboard

📱 Lite/Paid

Nice app that lets two people on the same Wi-Fi network work on the same piece of art simultaneously.

Draw Something

📱 Lite/Paid (+in-app purchase)

One of the fastest growing games ever (fifty million downloads in seven weeks), this collaborative drawing game has you sketching something and a remote player guessing what it is.

Engineering & design

AutoCAD WS

 Free

With this app installed you can easily view, edit and share DWG format drawings. Though some tasks can be quite frustrating, the sharing and collaboration tools work well, meaning you can deal with your designs without having to get to a laptop or desktop computer.

Home Design 3D

 Lite/Paid

This sophisticated application gives you the means to create accurate 2D and 3D plans of your home and then furnish it. Anyone familiar with Google SketchUp will feel right at home here. If you want to save your creations, you'll have to buy the Pro version.

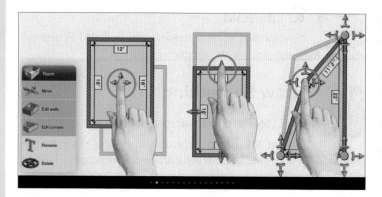

My Measures and Dimensions

Lite/Paid

This app allows you to take photos of a space, then overlay its dimensions (plus arrows and angles) and store the information in the image – incredibly useful for both professionals and DIYers.

Autodesk ForceEffect

Free

This is a very sophisticated engineering drawing app. As well as offering illustrative and annotating tools, it can simulate forces and weights to help you predict break points.

Intaglio Sketchpad

Lite/Paid

Though this app should arguably be avoided on the iPhone, it's great on the iPad. There's a useful toolbox of both freehand and geometric shape tools and grids and guides that do their job well. You can import and work on both vector (SVG) and rasterized (PDF, PNG, JPG) graphics.

Blueprint 3D

Paid (+in-app purchase)

Though technically a game, this extraordinary app will be of interest to anyone with an interest in graphic design. Solve the various jumbled puzzles and create challenges for your friends.

Printing from apps

The iPhone, iPod touch and iPad all support Apple's over-the-airwaves print technology AirPrint, which allows you to send documents, photos, webpages, etc, directly to your printer over a Wi-Fi network. Unfortunately, you have to own one of several pricey HP printers to make it work. For a full list of compatible machines, visit apple.com/uk/iphone/features/airprint.html. Once the new printer is set up, AirPrint printing is found by tapping the icon (in many apps), selecting the number of copies you want, and then tapping **Print**.

There are, however, many apps that offer printing functionality without the need to buy a fancy new printer.

Printopia

If you have a Mac computer with a printer connected to it, this is by far your best option. Once this Mac application is installed on your computer, your iPhone, iPod touch or iPad will recognize your printer in the same way it recognizes an AirPrint printer within the options menu.

What's more, the app also allows you to send the print job to your Mac as a JPG or PDF, or if you use Dropbox, directly to your Mac's Dropbox folder. Printopia can be downloaded from ecamm.com/mac/printopia.

PrinterShare2

Lite/Paid

This app lets you print photos, contacts' details, web-pages and the contents of the clipboard (i.e. anything clipped using the **Copy** command) directly from your device over the airwaves. For the full story, visit PrinterShare.com.

Utilities & file storage

Dropbox

Free (+subscription)

Dropbox is probably the best online file-storage system, making it a cinch to sync files and folders among multiple computers and mobile devices. It's free to use if you don't mind limiting your capacity to around 2GB, otherwise you might want to consider one of the various subscription packages. Both the iPad and iPhone interfaces of the app are intuitive and easy to use, though you're still sent back to the web for some sharing functions.

Google Drive

Free (+in-app purchase)

Many Gmail users are unaware that their Google account also gives them access to Google's cloud storage service, Google Drive. On an iPad and iPhone, this app is the best way in. Should you need it, you can use the in-app purchase option to subscribe to extra online storage space.

Air Display

📱 Paid

Turns your iPad or iPhone into an extra screen for your Mac. Use it, for example, to display a small-window application such as iChat while using your main Mac screen to get on with work.

Air Sharing

📱 Paid

This excellent universal app gives onboard file storage, Wi-Fi file sharing with desktop machines, access to networked printers and the ability to mount remote servers.

Jotnot Scanner

📱 Lite/Paid

A useful scanning tool. Use your iPad or iPhone's camera to take photos of the pages of a document and then store or share the resulting file. Shadows are automatically removed when the app processes the photos, so the results look more like the output of a flatbed scanner. The collection of images that make up the document can then be emailed as a single PDF or saved directly to iBooks, Dropbox, Evernote, Google Drive or whatever PDF reader or cloud storage app you have installed on your device.

And finally ...

Flashcards [+]

📱 Free (+in-app purchase)

Whether you are a student cramming for exams or an executive preparing a speech, this app is the one-stop-shop for penning easy-to-use flash cards. Aside from formatting tools, there are also multiple keyboards for accessing maths and chemistry characters. You can also go online and download other people's flash cards from both Quizlet and Flashcards Exchange.

Dragon Dictation

📱 Free

If you don't have a recent iPhone or iPad (which all have dictation tools built in), you'll love this impressive dictation tool. Simply tap the button and start talking, then edit with the keyboard if anything gets transcribed inaccurately. It's a great way to quickly draft long emails before copying and pasting into the Mail app.

VPN Express

📱 Free (+in-app purchase)

This is a very useful app when you find yourself needing to create a secure encrypted connection to the internet from your iPad or iPhone. The app is free, but you have to use in-app purchase to buy either your connection time or download capacity.

iThoughtsHD

▢ Paid (+in-app purchase)

A beautiful set of iPad tools for creating mind maps, with loads of export functions and the ability to sync with commonly used services such as Dropbox. An iPhone version is also available, but it's tricky to use on the smaller screen.

Mood Board

▢ Lite/Paid

Great for project planning, this app gives you a blank canvas on which to pin ideas, text, photos or whatever it is that gets your creative juices flowing. It's easy to use and excellent when working with colour palettes.

Diacarta Planner

▢ Paid

Falling somewhere between being a design experiment, a to-do list and a calendar, this app creates a design-heavy mind map of your plans for the day. It's a thing of beauty, though you're unlikely to find it replacing your regular planning tools any time soon.

Captio

▢ Paid

Genius app that fulfils the simple but very useful function of enabling you to quickly send yourself emails, notes, pictures etc.

16

Reference & Catalogues

What you need to know, when you need to know it

Web services (particularly the big search engines and Wikipedia) have made the acquisition of knowledge – though not always accurate knowledge – incredibly easy. But although you could go straight to Safari on your iPad or iPhone to search out all you need to know, there are plenty of specialist apps out there that both streamline the process and offer a few nifty twists. The Catalogues section of the App Store, meanwhile, is the place to head for niche reference, whether it be an app of 1000 tattoos or a guide to every breed of dog on the planet.

Dictionaries and general knowledge

Dictionary

 Lite/Paid

An easy-to-use and powerful dictionary that includes derivatives and British English spellings.

Dictionary.com

 **Lite/Paid**

Both a dictionary and thesaurus, this app has a well-designed interface, audio pronunciation and voice search, and much of the content can be accessed offline. The paid version is ad-free.

Longman Dictionary of Contemporary English

 Paid

Though not cheap, this app has good cross-referencing, making it a great choice when learning English as a foreign language.

Official Scrabble Words

Paid

This official Scrabble dictionary gives you access to all allowable words for the classic board game. It also has a built-in "solve" tool, for when you're stuck with a rack of nasties.

Wikipanion

 Lite/Paid

The free version of this app is excellent and delivers pretty much all you need to access Wikipedia in an app. The search tools are excellent and there are some nice reading preferences too.

Wiki Offline

 Lite/Paid

This well-designed app gives you access to Wikipedia's wisdom without an internet connection. The Lite version holds the 1000 most popular articles, while the paid app creates a local version of the entire website (which comes in at a surprisingly slight 3.5GB).

Wikihood

 Lite/Paid

Offering location-based access to Wikipedia, this app is great for finding out about buildings and sites in your local area. Be warned if you want to use it overseas though: it requires an internet connection and you could end up with some hefty roaming charges.

History Calendar for iPad

Paid (+in-app purchase)

This app tells you what happened on this day in history through an easy-to-navigate interface. The in-app purchase gives you unlimited access to the entries.

WordBook & WordBook XL

📱 📱 Paid

This app is another excellent dictionary and thesaurus combo, this time with the added ability to search for words with missing letters – making it great for solving crosswords.

Word Lookup

📱 📱 Lite/Paid

This is the best of the many anagram-solving dictionary-style apps. The paid version has no ads and can handle nine-letter words, while the free version only lets you go as far as seven letters.

Visual Thesaurus

📱 Paid

Putting a nice spin on the usual thesaurus app, this tool gives you an exploded mind-map-style presentation that you can drag and tap to dive deeper into the tree of synonyms.

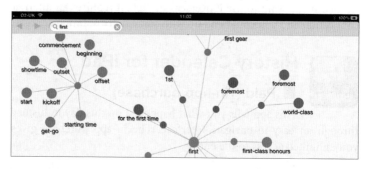

Space apps

Exoplanet

 **Free (+in-app purchase)**

Whatever you might be worrying about right now, this app should put things in perspective, with its complete overview of the vast night sky and the planetary objects therein. You can also receive a push notification whenever a new planet is discovered. In exchange for an in-app purchase you can kill the ads.

The Night Sky

 **Paid (+in-app purchase)**

This augmented reality app taps into your device's built-in GPS and compass gizmos when you look up at the night sky through your iPhone or iPad's camera. Your view of the sky is then overlaid with data about the stars, satellites and constellations that you see above you.

Moon Atlas

 Paid

Though you'll probably never actually go to the moon and find yourself needing directions, you might want some comprehensive lunar mapping on your iPad or iPhone … just in case. The mapping is presented on a manoeuvrable 3D globe and the level of detail is incredible. The same developers produce similar apps for several of the major planets, including Mars, Jupiter and Saturn.

Miscellaneous

It really does seem to be the case that any topic or subject you can think of is covered in the App Store somewhere, and while several of the selections below have snuck into this chapter from other, non-reference categories, they're all striving to be your one-stop shop for a particular area of knowledge – some with a real touch-screen twist.

Kings and Queens

 Paid

With its detailed timelines and family trees, this app is a content-rich romp through almost two thousand years' history of the British monarchy – great for swotting up on those all-important historical dates.

The Elements

 Paid

Bringing the periodic table kicking and screaming into the twenty-first century, this was one of the first apps to make a real splash when the iPad was launched. Its groundbreaking use of 3D photography and touch manipulation make it a pleasure to use.

Gems and Jewels

Paid

From the same developers that created The Elements, this iPad app presents a 3D exploration of precious stones.

SneakerFreak

🔲 Paid

On the off-chance that you desperately need a pocket guide to the most collectable trainers ever crafted ... this is it.

Design Icon Chairs

🔲 Paid

Part game, part reference tool, this app dishes up a slice of design history with information about styles, dates and makers. It features over one hundred classic chair designs from the nineteenth century to the present day.

Paper Sizes

🔲 Paid

Paper Sizes is a stylish little app that gives you a quick reference to the various standard paper sizes used in different parts of the world.

Tie Right

🔲 🔲 Paid

If you want to know the difference between a Windsor and a Half-Windsor or, more importantly, you need step-by-step instructions on how to tie one, this is the app for you. You can even adjust the on-screen instructions depending on whether you're looking "head on" into a mirror or from the "top down".

Ancestry

Free (+in-app purchase)

This beautifully designed UK-specific app allows you to create and edit your family tree wherever you are, adding images, notes and new discoveries with just a few taps. There is also a social element to the wider Ancestry.co.uk community (requiring a modest subscription charge).

Chirp! Bird Songs Europe +

Paid

One of the best bird reference apps in the Store (and there are many), this app contains hundreds of songs to help you recognize your little feathered friends when you hear them.

Flags and Capitals

Free

Great for quizzes and generally appearing to be a well-travelled individual; this app does exactly what you need it to.

Evi

Paid

Evi is an advanced "artificial intelligence" search tool that can be asked questions (via either voice or text) and, theoretically, will provide accurate, conversational responses. If you've enjoyed using Apple's Siri, download this and compare the results.

17
Social networking
Online community apps

Over the last few years there has been a massive shift in social networking activity from desktop computers to mobile devices, and with many people using sites such as Facebook as their primary means of contact with friends and family, it's hardly surprising that the shift has been so profound. Of course, apps have a significant part to play in all this, and you can find hundreds of tools and gizmos in the App Store that are designed to enhance your mobile social life.

Facebook

When you search "Facebook" in the App Store, you'll be confronted by an ocean of icons, each offering a different way into the omnipotent beast that is the world's most popular social network.

Facebook

Free

This is the "official" way to use Facebook on the iPhone or iPad. The app links well with the various notification and alert options available on your Apple device, meaning that you always know what new activity is going on in your social circle.

> **TIP** Under the Facebook main menu, tap **Nearby** to find places in your vicinity recommended by friends.

Aside from their main app, Facebook also offers several separate dedicated tools to streamline your experience. The Facebook Messenger client is worth looking at if you use Facebook to chat with friends while the Pages Manager app is aimed at those looking to optimize their brand or fan pages.

Facebook Messenger	**Facebook Pages Manager**	**Facebook Camera**	**Facebook Poke**
Social Networking	Business	Photo & Video	Social Networking
Free ▾	Free ▾	Free ▾	Free ▾

ChatNow for Facebook

Free

Similar to Facebook's own Messenger app, this friendly chat client for Facebook has nice backgrounds and fonts, and comes with built-in emoticons.

MyPad – For Facebook and Twitter

Lite/Paid (+in-app purchase)

The iPad version of this app is particularly impressive, and the ability to switch between Facebook and Twitter feeds in one place is both useful and intuitively tackled.

Cover Photo Maker for Facebook

Lite/Paid (+in-app purchase)

This nice set of tools can be used to create custom Facebook Cover Photos for your account page. The in-app purchase unlocks extra fonts and upgrades you to the ad-free Pro version.

Beejive for Facebook

Free

Another excellent chat and messaging client for Facebook, this time from Beejive, one of the most popular providers of mobile messaging services. The built-in toolkit of features is well laid out and includes very good notification options.

Twitter

No doubt you have a fairly good idea of what Twitter is all about. As with Facebook, it has really blossomed thanks to mobile technology.

Twitter

📱 Free

Most Twitter clients do pretty much the same thing, so it's down to personal preference as to how you want your tools laid out. The popular official choice is well designed, allows for multiple accounts and boasts real-time search and a handy trending view.

Tweetbot

📱📱 Paid

Dubbing itself a "Twitter client with a personality", Tweetbot is clean and elegant, with an excellent set of tools.

Trickle for Twitter

📱 Paid

This app offers passive, easy-on-the-eye tweet consumption without all the additional bells and whistles.

Twittelator Neue

📱 Paid

Specifically designed afresh for the iOS 5 operating system, this colourful Twitter client is well worth checking in with.

Miscellaneous

LinkedIn

 Free

A well-built app worth installing if you're a regular user of the popular professional networking service. They also make CardMunch that snaps business cards and adds them to Contacts.

Tumblr

Free

Tumblr is a first-rate blogging and social media platform, and this app is an essential piece of your mobile Tumblr kit. You can use it to keep track of all your blogging activity, to post both text and media, and to respond to queries from your readers.

Google+

Free

Google's social network is yet to achieve the traction of Facebook, but the app is solid. The Google Hangouts conference chatting is particularly well handled.

Pin HD for Pinterest

Lite/Paid

The Pinterest virtual pinboard network has its own official app, but this universal app client gets our vote.

AudioBoo 2

📱 Free

If you think audio posting "Boos" is going to be the next big thing, download this app and join the craze.

Digisocial

📱 Free

Combine pictures and audio clips and then post them to the network. You can also share via Facebook and Twitter and send instant voice messages to your friends.

Keek

📱 Free

Though it has some way to go before rivalling Facebook and YouTube, this video-posting app is one to watch.

Cirqit

📱 Paid

One of many apps now available that allows you to quickly connect with friends based on their location.

TIP For other ways to communicate via apps, turn to the Utilities chapter for chat, voice and video-calling tools.

18

Sports

Having a ball on your iPad or iPhone

You can seemingly find any sport in the App Store, but some are better represented than others. Soccer, golf and cycling account for a large portion of the apps on offer, with the remainder mainly taken up with hunting calls and fishing advice. If you want to follow a specific team or get live scores for a particular sport, you're almost certainly in luck; just search the store to find what you're looking for. Meanwhile, here's a selection of the most interesting sports apps we could dig up.

American football

American Football - Understanding The Game

 Paid (+in-app purchase)

For those who play the sport, this pocket guide to the rules of American football would be a useful addition to your app library. For Brits trying to watch the sport and make sense of it, this isn't just useful reading – it's essential.

Cricket

ESPN Cricket Scores & News

Free

This comprehensive pair of world cricket apps for the iPhone and iPad dish up everything from scores and fixture listings to live coverage and journalistic podcasts. Sky Sports produce a similar app that's also worth investigating.

Cricbuzz

Free

Another good choice for aggregated scores and player profiles, this cricket app has a really nice interface, and the ball-by-ball commentaries are excellent. It also delivers push notifications every five overs or whenever a wicket falls.

Cycling

Cyclepedia

 Paid

This beautiful app gives you a whistle-stop tour of the history of cycle technology. There are various ways to navigate your way around the content, though the timeline view is a real treat. Once you start exploring the individual cycles you discover that there are 3D views and stunning magnified views that really let you explore the detail.

CoachMyRide

 Paid

Well-designed cycle training app, with structured pro-grammes written by elite coach Lionel Reynaud.

Bike Repair

 Paid

A step-by-step bicycle repair and maintenance guide that covers more than seventy different problems and how to fix them, in an easily navigable interface. For more, try the iPhone-only app Bike Doctor.

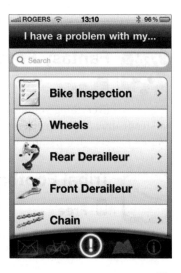

Soccer

Live Scores Sofa Score

 Free

Get real-time results from around the world direct to your iPhone or iPad. This app covers tennis, basketball, handball, Formula One, baseball, rugby and ice hockey, as well as soccer. For more scores coverage, try ScoreMobile, or for soccer only, try Soccer Livescores or Goal.com Mobile.

ESPN Goals

 Free

This app reports on all the goals news as it happens in the UK Premier League.

Fantasy Football Manager (FFM)

 Paid

Everything you need to manage your own fantasy football team: drag and drop players into position, get the latest news on injured players, view fixtures and sort teams and players.

Tribal Football

Free

Must-have app for all the latest soccer fixtures and transfer news, along with live scores and curated Twitter feeds.

Golf

Golfshot: Golf GPS

Paid

This pricey but excellent GPS-based golf range finder shows your distance from the green or to any other point on the course. It includes satellite maps and detailed scorecards for countless golf courses, and if you find a course that's not covered, you can request it be added to the database. If you need to improve your play (and let's face it, who doesn't), the same developers have produced Golfplan with Paul Azinger, which is loaded with tutorial videos and useful tips.

FreeCaddie

Lite/Paid

If you need even more golf GPS in your life, then try out FreeCaddie. The app covers 21,000 courses and has an interface that is easier to read in the sunshine than that found on many similar products. If you want additional info about hazards and bunkers, you'll need to shell out for the Pro version.

Golf It, Score It

Paid

With this app, you can keep score for up to four players at a time. There are several useful charting and statistics tools, plus the option of emailing your scorecards.

Motor sports

iRally

 Free

By far the best app in the Store for following the goings-on in the world of rally driving, covering both IRC and WRC.

F1™ 2013 Timing App

📱 **Free (+in-app purchase)**

Though not as pricey as actually going to watch F1, let alone buying the T-shirt to show you were there, this app's "championship pass" in-app purchase doesn't come cheap. It does, however, offer excellent track and event information, live track positions and all sorts of other useful info.

Skiing

Ski & Snow Report

📱 Free

One to check before you hit the slopes, this app offers ski reports, five-day weather forecasts, live cameras and first-hand accounts from other skiers, all in the context of a very nice, easy-to-read design.

SkiMaps

📱 Paid

Covering the most popular resorts across Europe and Canada, this is an app of GPS-enabled piste maps. Also thrown in is the ability to record data about your performance on different runs, and best of all you can do all this without an active internet connection (though be sure to download the maps you want before you head out the door).

Miscellaneous

CoachNote

📱 Lite/Paid

This very well designed app allows you to create strategy diagrams for planning your game in everything from football and cricket to ice hockey and lacrosse.

Simple Interval Timer

📱 **Free**

As it says, this is a straightforward and user-friendly timer to be used during any kind of training or gym session. It can run in the background, meaning you can use your phone to play music or watch videos while you train.

Scoreboard

📱 **Paid**

Complete with themes, shot clocks, time buzzers and reset gestures, this app is all you need to keep track with the action on the field … whichever sport you are watching or umpiring. If you are after a free alternative, try FlipScore.

Team Stream (HD)

📱 📱 **Free**

This Bleacher Report app is justifiably well-respected. Giving you access to news, scores and videos from your favourite teams, it covers a variety of sports, including soccer, American football and basketball.

19
Travel
Getting from A to B via the App Store

Travel is all about going places, and when it comes to apps that dish up information about those places, there's no shortage of choice in the App Store. But be careful when choosing travel guide and destination apps, as many do little more than pull articles from Wikipedia. As for actual transportation, check to see whether your local train or bus company has a dedicated app – they almost certainly will, and it will almost certainly become an indispensable travel companion. In the UK, MyNet Travel and UK Train Times are essential downloads for anyone travelling by rail; in the US, try either the official Amtrak app or the excellent RailBandit.

General travel

The main drawback of travelling with an iPhone or iPad is the issue of roaming charges, so if you download guides before you set off, make sure they include offline content and maps (as is the case with the Rough Guides travel apps), meaning you can use them with data roaming turned off (**Settings > General > Mobile Data**).

Rough Guides Travel Survival Kit

Paid

This app covers everything from how to avoid getting mugged to what to do if you come face to face with a bear. It also has a useful section for storing visa and passport numbers, as well as quick-dial links to local embassies and your next of kin.

Tripit

Free/Paid (+Pro subscription)

A useful app that aggregates all your trip information (flights, transfers, hotel bookings etc) in one place. The paid app comes sans ads, while the addition of a Pro account gives you live flight updates amongst other perks.

Pin Drop

Free

Falling somewhere between being a social network and a constantly evolving personalized travel guide, this app lets you bookmark locations, view others' pins and much more.

Flights

FlightTrack

🔲 Lite/Pro (+in-app purchase)

This app is a revelation for frequent flyers. It gives you all the live flight info you could wish for, syncs your bookmarked flights to your calendar and can be used to track the status of any flight, either on the ground or in the air. You can also overlay a weather feed to the flight maps to see where your ride might get a little bumpy, and there's an offline mode you can use while in the air. Dig down into **Settings** and you'll find options for syncing with iCloud and the aforementioned Tripit.

Skyscanner

📱 **Free**

A great app for finding the cheapest flights. It searches across nearly all the airlines and resellers and serves up the results in an easily digestible form.

British Airways

📱 **Free**

Most airlines now have their own app; BA's lets you manage your flights and air miles, check in before you reach the airport and use your iPhone as a paperless boarding pass (at certain airports).

Airport Guide – iFLy

📱 📱 **Lite/Paid**

An extremely useful app featuring guides to hundreds of airports across the US, including shopping and restaurant maps, baggage allowance data and travel tips. Many specific airports have similar apps, so check the Store before you travel.

Plane Finder AR

📱 **Lite/Paid**

This extraordinary app allows you to point your phone at any plane in the sky to find out all sorts of information about it. The same company also makes the Ship Finder AR app.

Hotels

Jetsetter

 Free

In contrast to the clutter of most online booking experiences, this app is all about presenting you with the best hotels within a sumptuous and uncluttered digital arena. With 360° panoramas (on the iPad version), loads of photography and reviews, this app makes picking your hotel almost as fun as actually going there.

Hotels.com HD

 Free

Most of the big booking sites have apps, but this iPad-only app is a real treat, with an interactive city skyline tool and thousands of last-minute deals.

Hostel Hero

 Free

This essential backpackers' tool recommends thousands of hostels and budget hotels across the world.

Mr & Mrs Smith: Plan & Play

 Free

A very nicely designed app dishing up boutique hotel listings, with a cheeky game thrown in too.

Language & phrases

Word Lens

 Free (+in-app purchase)

Point the camera of your iPhone or iPad at a foreign sign and this app will translate it into your native language. Although such technology is still in its infancy, it's extraordinary to use.

Google Translate

 Free

This Google tool translates spoken or typed text in dozens of languages. The translated phrase can be displayed in an extra-large landscape mode – handy for getting your message across.

SayHi Translate

 Paid

Another excellent translation app with voice recognition and impressively accurate results. As well as covering forty languages and dialects it has some nice sharing tools.

iTranslate+

 Lite/Paid (+in-app purchase)

This app allows for spoken translations and boasts a wide selection of languages. It also has a conversation mode, integrated dictionaries and the ability to auto-detect some dialects.

Miscellaneous

 ## Gogobot – Travel Postcards

📱 Free

This app gives you the tools to create postcards to share with friends and the wider Gogobot community while you're travelling. You can also use it to explore places and things to do nearby and to read other people's recommendations.

 ## City Bikes

📱 Free

Essential download for anyone who wants to use a city bike scheme. This app shows you all the bike stations across your metropolis, including how many spaces are free at each one. The app also has a built-in timer so that you don't miss your bike return time. There are dozens of cities currently covered, include Washington, Paris, Toronto, London, San Francisco and Barcelona.

 ## Rough Guides Trip Lens

📱 Free

For travel journaling on the iPhone, this Rough Guides app allows you to organize your posts, photos and destinations within a travel timeline. The app is intuitive and features a very cool "Auto" mode, whereby your phone records your GPS activity in the background while you get on with being a tourist. Another good travel journaling tool worth looking at is Off Exploring.

World Lens

Free

This is Rough Guides' home-grown app of inspirational travel photography, with built-in tools for bookmarking and sharing images via email, Facebook and Twitter.

National Trust

Free

For anyone travelling around the UK or just looking for a nice day out, the official National Trust app is a great starting point. All the sites are map located, and there is plenty of detailed visitor information, plus photos, to help you make your choice.

Stow

Paid

Nice app for creating packing lists for your trips. With templates and weather-specific packing prompts, this app takes the pain out of filling a suitcase.

Trip Wallet

Paid

A handy little app for keeping track of your holiday spending, with graphs and budget tools.

20
Utilities

You'll wonder how you ever lived without them

As with many of the categories in the App Store, the Utilities section is a mixed bag of the obvious, the ridiculous and the just plain awful. On the whole, you can expect to find some very functional products that more often than not fall into one of the sub-categories defined in this chapter: web browser apps, email apps, calling apps, and so on.

Web browsers

There are several browsers in the App Store, but none offers all the features and integration of the built-in option: Safari. If you want to try something new, search both the Utilities and Productivity categories to see what's available, or check out one of these:

Chrome

 Free

The iOS version of Google's flagship web browser is just as slick and function-rich as its desktop cousin. Best of all, it's really fast to load pages, and the tabbed browsing works better than Safari page setup. Logging in with your Google account allows you to sync tabs and bookmarks with your other computers and devices.

Opera Mini

 Free

From the same people who make the brilliant desktop Opera browser, this iPhone and iPad version is impressively fast and features tabbed browsing, password tools and a quick access "Speed Dial" tool to get you to your favourite sites.

Mercury Web Browser

 Lite/Paid

Another very handy Safari alternative. This one can be decorated with themes, features Firefox syncing tools and Dropbox integration, and has various handy finger gestures for swift browsing.

FREE Full Screen Private Browsing for iPhone & iPad

 Free

A no-frills affair that offers completely private browsing sessions: no history, no cache, no cookies etc. Although you can achieve the same thing by enabling the function in Safari (**Settings > Safari > Private Browsing**), you might find it easier to have a separate app available that always behaves in this way. Catchy app name too!

Email clients

Sparrow

Paid

This elegant alternative email client offers some really nice features over and above those found in Mail on the iPhone. You can have contacts' images appear on your messages and threads, swipe easily through conversations and between accounts and, best of all, drag from the top of the screen to refresh your inbox.

Gmail

 Free

Google's Gmail app is fast, easy-on-the-eye and has a very nice contrasting sidebar menu that supports coloured labels. If you use multiple Gmail accounts, you might also want to look at the feature-rich Safe Gmail app.

Calendar apps

There are hundreds of calendar apps in the store – from apps created by specific sports teams and TV shows to apps for keeping track of lunar, solar and even fertility cycles. So if you're interested in anything date-specific, search the App Store to see what's on offer.

Agenda Calendar

 Paid

This calendar app is minimal and fully featured, with a nice year-long "goal" view. The app integrates well with Notification Center and features some handy gestures for ease of use.

CalenGoo

 Paid

A popular, colourful app that syncs all your Google account calendars and offers a wide range of views. This app also boasts "floating events", which are like a mix of events and to-do items, with a check box to tick when completed. You can also attach documents to individual events and sync them via Google Drive.

Calendars – Google Calendars

 Paid

This fully featured alternative calendar for your iOS device offers drag-and-drop event editing and SMS alerts. You can also zoom in on the month view, making it easy to target individual events and see details.

iPlanner UK

📱 **Paid**

This nicely put-together app gives you loads of flexibility (and some chirpy little icons) when creating events within its landscape-only view. The best thing about iPlanner, however, is the full-year view, which many other calendars don't have.

Big Day – Event Countdown

📱 **Lite/Paid**

Whether it's an upcoming wedding or the beginning of your summer vacation, this joyous little app will help you count down to the big event.

Calvetica Calendar

📱 **Paid**

Feeding off the regular accounts you might already have set up on your iPad or iPhone, this calendar app has a good set of tools under its belt and looks gorgeous.

Grid Diary

📱 **Paid**

Though not technically a calendar app, this diary-keeping app has a really nice calendar-like view and a unique approach to journaling. The app asks you questions each day about what you've done in order to help you compose your entries.

Calling apps

Dial Plate

 Paid

This is one of many apps that add a retro rotary dialler to your iPhone's armoury.

A Fake Caller

 Lite/Paid

A fun way to schedule "fake" incoming calls, and useful if you want to impress your friends by having the president phone – or if you need an excuse to get away from an awkward social situation.

Face Dialer

Free (+in-app purchase)

A handy app for creating picture shortcut icon buttons for commonly used contacts. There are dozens of similar apps available: Call Him and Call Her are among the more stylish offerings for speedy spouse dialling.

calLog

 Paid

When used in place of the iPhone's built-in call list interface, this app allows you to add notes and reminders to your calls and also to export reports on call activity.

0870

 Free

Mobile-phone users in the UK get stung by high calling rates to 0870, 0800, 0844 and 0845 numbers. This app provides a conventional landline substitute for each number.

Record Phone Calls

Paid

One of several apps offering the ability to record phone calls. Handy, though for ethical (and sometimes legal) reasons, it's important to tell the other person that they're being recorded. This app will record up to 45 minutes of conversation at any one time.

ChatTime

Free (+in-app purchase)

This allows you to make inexpensive international calls via the regular phone network, rather than calling via the internet.

HulloMail SmartVoicemail

Paid (+in-app purchase)

This advanced cloud-based visual voicemail service allows custom messages for different callers, text transcriptions of the messages people have left for you, and the ability to share messages with your Evernote account (see p.116). Use the in-app purchases to buy extra features and subscribe.

Calling via the internet

Anyone accustomed to using Skype will be aware that it's possible to make free or virtually free internet calls – voice and video – to land-lines and mobiles all over the world. Here are some of the options:

Skype

🔲 🔲 **Free (+in-app purchase)**

This app gives you access to most of the same features you get in the computer version of Skype: instant messaging, Skype-to-Skype calls and, if you buy some credit, very inexpensive Skype-to-phone calls. Skype-to-Skype calls can be made for free via Wi-Fi, 4G or 3G – though bear in mind that you'll chomping through your data allowance when using cellular networks.

> **TIP** Skype also makes the very handy Skype WiFi app, which lets you use your Skype credit to join thousands of paid Wi-Fi networks around the world.

FaceTime

 Built-in

Assuming FaceTime is switched on (you can check within **Settings > FaceTime**), making a video call to other FaceTime users is as easy as tapping a button – though you'll need to be on Wi-Fi, 4G or 3G to do it. You can contact people via either an iPhone mobile number or their Apple ID email address. And because the calls go over the internet rather than the phone network, they're free (except for any data charges if using 4G or 3G).

Truphone

📱 Free (+in-app purchase)

This very good alternative to the Skype app gives you VoIP call contact with Skype, Truphone and Google Talk users, plus a free voicemail service with push notifications. It works really well over Wi-Fi, 4G and 3G and can run in the background.

Fring

📱 Free (+in-app purchase)

Like Truphone, Fring gives you access to contacts on a range of networks, this time including Windows Live Messenger, Google Talk, Twitter, Yahoo!, AIM and ICQ. The service's main offering, however, is its ability to support four-way video calls – a bit like a conference call version of FaceTime.

Hang w/

📱 Free

With another twist on the streaming video theme, this app has you following people (virtually) and then receiving a notification when they are online broadcasting live and ready to chat.

Vtok

📱 Free (+in-app purchase)

For Google users, this app gives you no-nonsense access to voice, video and chat services.

Messaging apps

Emoji>

Free (+in-app purchase)

If you want to use emoticons, also known as emoji icons, in your messages and emails, you could download an app for the purpose. This one is a good choice. Once the app is installed, navigate to **Settings > General > Keyboard > Keyboards > Add New Keyboard** and enable the **Emoji** option. From then on, the new keyboard is available via the keyboard-switching "globe" button to the left of the spacebar on your keyboard.

WhatsApp Messenger

Paid

A good alternative to Apple's iOS-only iMessage service, this cross-platform SMS-style messaging service is available for all smartphones and works over the internet, meaning you don't get charged per-message, as you might with your traditional SMS quota.

BeejiveIM

Paid

Beejive is another great chat service app, which allows you to chat simultaneously across a range of services (Yahoo!, GoogleTalk, Facebook, MySpace, Jabber, etc). It's one of the best clients when it comes to getting notifications on your iPhone or iPad, meaning you always know when people are trying to contact you or have left you a message.

Time apps

Alarm Clock

📱 **Lite/Paid**

The App Store has plenty of choice on offer if you are looking for a half-decent night-stand alarm-clock app. This one features access to your device's music library along with podcasts and audiobooks. The live weather feed is a nice addition.

TimeTuner Radio Alarm Clock

📱 **Paid**

A full-featured and stylish bedside alarm clock and internet radio tuner, with unlimited alarms and a snooze function.

Alarm Clock HD

📱 **Paid**

This is a really well put-together night-stand-style alarm-clock app, with a Google Reader feed ticker and music sleep mode.

Flick Clock

📱 **Paid**

Not the most exciting app in the world, but if you like that retro vibe of a flip clock on your desk, download it. A free alternative, though not quite so stylish, is Flipclock.

Nooka

 Lite/Paid

Though arguably more of an interesting design experiment than a particularly useful timepiece, Nooka takes all things clock-related (global time zones, stopwatches, timers, alarms etc) and presents them within a series of visually striking formats.

World Clock Pro

 Paid

This universal app is a great addition to any iPad or iPhone app collection. The interface is lush and makes it easy to add multiple cities to the view. There are several custom themes and a very useful split-screen world map view.

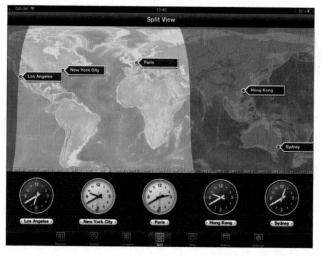

Security apps

Lockitron

Free

This app offers a glimpse into the near future as it enables your iOS device to become a key for opening doors … assuming you're intending to open a door with a Lockitron electronic lock, that is. It works via a server that lets your door know to either lock or unlock. You can also use the app to issue unlock privileges to friends and neighbours, and review logs of when doors have been used. Check out the Lockitron website for details about the associated hardware (currently only available in the US).

Dot Lock Secure

Lite/Paid (+in-app purchase)

Use this app to secure passwords, contacts, notes and the like behind a special pattern password screen – the dot lock! You also get intruder reports (including the culprit's photo, snapped by your front-facing camera) should someone try to open the app and fail.

1Password

Paid

A password manager that stores all your log-in details within an encrypted database on your iPad or iPhone. Though it doesn't hook in with Safari, its log-in tools can be used within its own integrated browser – which might be worth using for particularly sensitive websites.

Miscellaneous

AirPort Utility

 Free

This is a very useful app for tweaking the settings of Apple AirPort base station networks.

Universal Converter

 **Lite/Paid**

The only iPad conversion tool you need: it covers weights, measures, forces etc. The in-app purchase adds currencies.

> **TIP** If you have a recent iPhone or iPad, you can also try Siri for voice-activated unit conversions.

Decibel Meter Pro

 Paid

How loud can you shout? Or more to the point, how loud are your neighbours shouting?

Data Usage

 Lite/Paid (+in-app purchase)

Get detailed stats and information about the data you're consuming over your cellular phone network.

Clinometer

📱 **Lite/Paid**

There are loads of apps that can act as a spirit level, but this one is particularly well thought through and easy to read.

Speedometer

📱 **Paid**

Though there are free alternatives in the Store that will tell you how fast you are going, this paid app has a clear interface and a useful "reflective" mode so that you can see the readout on your windscreen when your phone is on the dashboard.

FlickKey Keyboard Notes

📱 **Paid**

This note-taking tool features a revolutionary nine-key keyboard. To use any of the nine keys, tap and then flick your finger in the direction of the letter you want. As you can imagine, it is infuriating to use at first, but remarkably intuitive once you get the hang of it.

Speedtest X HD

📱 **Lite/Paid**

This very useful app helps you test the speed of the Wi-Fi or cellular network that you are currently connected to, which can be particularly useful when trying to diagnose a connection issue.

Onavo

 Free

This incredibly useful app (and associated cloud service) runs in the background, crunching away at the data you download via your iPhone's network data plan. Using some very clever compression techniques it reduces the amount of information that passes through your data quota and can end up saving you a lot of cash, or at least making sure you get the most value out of your data allowance. It can be especially useful when you're abroad.

RedLaser

 Free

An essential app for reading barcodes and QR codes (such as the one shown here) when you're out and about. For product barcodes, it will give you an instant readout of all the online stores where you might choose to buy the item (including prices). QR codes, meanwhile, are often used on posters or in magazines to quickly take you to a webpage.

Tally-Ho!

 Paid

A useful app for counting stuff. You can manage multiple tallys, it features password protection, and the interface is very clear and easy to use. For a free alternative, try Tally It!

21
Weather
Apps for a rainy day

The iPhone's built-in Weather application does pretty much what you would expect; you can save favourite locations and view current conditions along with a six-day forecast. For a quick update of the weather, you can also look to the Notification Center, or even ask Siri. But the built-in experience isn't to everyone's taste, and you may well have an alternate weather source that you trust over Yahoo! (which Apple use for their weather data). And then there are the iPad owners, who don't get the Apple Weather app built in at all. Here are a few recommendations to help you prepare for whatever Mother Nature might have to throw at you. Or you could just look out the window ...

Weather apps

Met Office Weather application

 Free

For the UK, the Met Office alternative to the Apple Weather app is a must, largely because it includes radar and satellite imagery. Also, the five-day forecast tends to be more accurate.

Weather Live

 Paid

Claiming to be "the most beautiful weather app", Weather Live certainly has a good stab at the title. The animations are delightful, and there are loads of destination search filters and weather parameters to play with.

Fahrenheit & Celsius

Lite/Paid

These two separate apps keep you up to speed with the current temperature, and handily display it as a badge icon.

Weatherwise

Free (+in-app purchase)

Another weather-feed app with a twist: this time the location-specific data you're after is presented alongside some seemingly random custom artwork themes.

Weddar

📱 **Free**

Dubbed "Twitter for weather" by *Wired* magazine, this app puts weather reporting in the hands of the masses and allows you to tune into what they're saying. It's an interesting concept, and the app is fun to use, but expect to come across some pretty wild interpretations of what actually qualifies as weather reporting.

Weathercube

📱 **Paid**

This unusual weather app presents the data in the form of a cube that can be pulled around and expended using some nice finger gestures. Within the Settings you can set specific locations and change the colour scheme.

Magical Weather

📱 **Paid**

By far the best-looking weather app for the iPad, this software allows for up to eight saved forecast locations.

WeatherMap+

 Paid

The detail given on these interactive maps is amazing, allowing you to switch between layers for temperature, wind speed and pressure, giving you a real sense of global weather changes.

Netatmo

Free (+hardware)

If you own a Netatmo weather station, then this accompanying app gives you access to your accumulated data and observations. Find out more about the device at www.netatmo.com.

Hurricane Track for iOS

Free

Offering live radar feeds, hurricane path predictions and cloud analysis data, this app is an essential download for all hurricane watchers – and those that live in their potential path.

Quakes – Earthquake Notifications

 Free

With push notifications coming from sites around the world, as and when earthquakes happen, this app can be a frightening companion. The data is fascinating, however, and in-app purchases grant you additional functions and no ads. Also checkout QuakeWatch.

Index